NANOCRYSTALLINE APATITE-BASED BIOMATERIALS

NANOCRYSTALLINE APATITE-BASED BIOMATERIALS

D. EICHERT
C. DROUET
H. SFIHIA
C. REY
AND
C. COMBES

Nova Science Publishers, Inc.
New York

Copyright © 2009 by Nova Science Publishers, Inc.

All rights reserved. No part of this book may be reproduced, stored in a retrieval system or transmitted in any form or by any means: electronic, electrostatic, magnetic, tape, mechanical photocopying, recording or otherwise without the written permission of the Publisher.

For permission to use material from this book please contact us:
Telephone 631-231-7269; Fax 631-231-8175
Web Site: http://www.novapublishers.com

NOTICE TO THE READER

The Publisher has taken reasonable care in the preparation of this book, but makes no expressed or implied warranty of any kind and assumes no responsibility for any errors or omissions. No liability is assumed for incidental or consequential damages in connection with or arising out of information contained in this book. The Publisher shall not be liable for any special, consequential, or exemplary damages resulting, in whole or in part, from the readers' use of, or reliance upon, this material.

Independent verification should be sought for any data, advice or recommendations contained in this book. In addition, no responsibility is assumed by the publisher for any injury and/or damage to persons or property arising from any methods, products, instructions, ideas or otherwise contained in this publication.

This publication is designed to provide accurate and authoritative information with regard to the subject matter covered herein. It is sold with the clear understanding that the Publisher is not engaged in rendering legal or any other professional services. If legal or any other expert assistance is required, the services of a competent person should be sought. FROM A DECLARATION OF PARTICIPANTS JOINTLY ADOPTED BY A COMMITTEE OF THE AMERICAN BAR ASSOCIATION AND A COMMITTEE OF PUBLISHERS.

LIBRARY OF CONGRESS CATALOGING-IN-PUBLICATION DATA

Nanocrystalline apatite-based biomaterials / D. Eichert ... [et al.].
  p. ; cm.
Includes index.
ISBN 978-1-60692-080-0 (softcover)
1. Bone substitutes. 2. Nanocrystals. 3. Apatite. I. Eichert, D.
[DNLM: 1. Bone Substitutes--metabolism. 2. Apatites--chemistry. 3. Biomimetic Materials--chemistry. WE 200 N186 2009]
RD755.6.N36 2009
610.28--dc22
                        2008042003

Published by Nova Science Publishers, Inc. ✢ *New York*

CONTENTS

Preface		vii
Chapter 1	Introduction	1
Chapter 2	Early Works and the Way Bone Mineral was Conceived	3
Chapter 3	Synthesis of Nanocrystalline Apatites	9
Chapter 4	Characterization of Apatites	13
Chapter 5	A Model for Nanocrystalline Apatites	29
Chapter 6	Physico-Chemical Properties of Nanocrystalline Apatites	33
Chapter 7	Processing of Nanocrystalline Apatite-Based Biomaterials	45
Chapter 8	Biological Properties of Nanocrystalline Apatites	59
Conclusion		63
References		65
Index		77

PREFACE

The improvement of the biological activity and performance of bone substitute materials is one of the main concerns of orthopaedic and dental surgery specialists. Biomimetic nanocrystalline apatites exhibit enhanced and tunable reactivity as well as original surface properties related to their composition and mode of formation. Synthetic nanocrystalline apatites analogous to bone mineral can be easily prepared in aqueous media and one of their most interesting characteristics is the existence of a hydrated surface layer containing labile ionic species. Ion exchange and macromolecule adsorption processes can easily and rapidly take place due to strong interactions with the surrounding fluids. The ion mobility in the hydrated layer allows direct crystal-crystal or crystal-substrate bonding. The fine characterization of these very reactive nanocrystals is essential and can be accomplished with different tools including chemical analysis and spectroscopic techniques such as FTIR, Raman and solid state NMR. The reactivity of the hydrated layer of apatite nanocrystals offers material scientists and medical engineers extensive possibilities for the design of biomaterials with improved bioactivity using unconventional processing. Indeed apatitic biomaterials can be processed at low temperature which preserves their surface reactivity and biological properties. They can also be associated in various ways with active molecules and/or ions. Several examples of use and processing of nanocrystalline apatites involved in the preparation of tissue-engineered biomaterials, cements, ceramics, composites and coatings on metal prostheses are presented.

Chapter 1

INTRODUCTION

Osteoarticular pathologies are, at all ages, the first cause of handicap and raise concern in public healthcare. Articular aging, traumatology, child growth defects, bone tumor treatment, osteoporosis and related bone failures, constitute points for which research efforts are necessary. Bone diseases and induced defects are also of prime importance in maxillo-facial surgery and odontology especially for aging populations. The shortcomings of bone auto- or allografts which in addition involve secondary operations, risks of disease transmission as well as immunological rejection and morbidity justified the development of synthetic bone graft materials.

In the past decades, the effort to adapt the first biomaterials taken from other technical domains (e.g. alumina, carbon, titanium carbide and nitride, plaster of Paris), or to design new materials better suited to biological applications, led to significant advances. However in the opinion of many researchers the ultimate achievement would be the perfect imitation of biological tissues and more importantly the improvement of biological repair and maintenance processes. Ideally, a substitute material should mimic the living tissue's mechanical, chemical, biological and functional properties; however the design of a complex structure such as bone is still impossible to achieve without the aid of Nature which masters, using sophisticated chemical properties and processes, this high performance mineral-protein composite.

Poorly crystalline apatites (PCA) are the major inorganic constituent of mineralized tissues in vertebrates. The imitation of bone mineral has inspired the research and development of calcium phosphate (CaP) based biomaterials. The most famous and most widely used CaP compound is stoichiometric hydroxyapatite processed as dense or porous ceramics, coatings and composites.

However, the design of biomaterials has evolved and today the function of a biomaterial is not restricted to physical substitution but the novel biomaterial should actively participate in the process of bone regeneration implying reactivity of the component(s). In this view we will show in this chapter how non-stoichiometric nanocrystalline apatite-based biomaterials can fulfil two of the main challenges: mimicking bone mineral crystal structure and composition and exhibiting a controlled reactivity regarding interactions with components of biological fluids (ions, proteins).

This chapter reports part of our "bioinspired" research based on the idea that the development and the processing of bioactive biomaterials for bone substitution or regeneration applications can take advantage of a thorough knowledge and understanding of the structure and properties of the tissue to be substituted. Due to the complex structure and heterogeneity of biological systems such as bone, synthetic apatites are generally used to throw light on the surface reactivity of bone mineral. However the characterization of nanocrystalline apatite analogous to bone mineral is difficult due to its relative instability and poor crystallinity. Even though the results are complex, recent fine investigations on synthetic PCA revealing their original surface reactivity are useful both in biomaterials or biomineralization.

The present chapter gathers the main results of our most recent investigations on nanocrystalline apatite composition, structure and properties with the aim of better understanding bone mineral properties and achieving better design of new bioactive nanocrystalline apatite based biomaterials or improving the biological performance of existing bone substitutes. This chapter which alternates between results on synthetic apatites and their significance for biological apatites is organized in seven sub-sections: 1) a description of bone composition and structure and the way it was perceived through the 20th century are presented first, 2) synthetic routes of nanocrystalline apatites are discussed with an emphasis on biomimetic nanocrystalline apatite preparation and maturation at room temperature and at physiological pH as developed in our research group, 3) the complementary characterization techniques that we use to investigate the global and local fine structure and composition of nanocrystalline apatites are reported 4) a model for apatite nanocrystals is put forward which can explain the properties of synthetic and biological apatite nanocrystals, 5) the physico-chemical properties of biomimetic apatites and their involvement in the biological behavior of PCA based materials are presented and discussed, 6) examples of nanocrystalline apatite based biomaterial showing some potential with regards to nanocrystalline apatite surface reactivity are described, 7) finally, the biological properties of apatites are presented.

Chapter 2

EARLY WORKS AND THE WAY BONE MINERAL WAS CONCEIVED

The studies on bone mineral composition and structure were rather puzzling for the first investigators. Before the development of structural analysis by X-ray diffraction (XRD), until the beginning of the 20th century, chemical composition was one of the major characterization tools used to identify biominerals. Other identification methods, for example based on the use of the optical properties of crystals could not be applied to biological apatites due to the small size of their crystals. Chemical analyses revealed the diversity of phosphate-containing biominerals. Three major components are always present: calcium, phosphate and carbonate, and were readily identified but they showed variable contents depending on the species, the individuals, their location in the body or the age and the type of mineralized tissue considered, in contrast with the other major biomineral, calcium carbonate, showing a rather constant composition. It was then accepted, in accordance with the hypothesis of Haüy concerning the composition of calcium phosphate-carbonate minerals, that several phases co-existed in hard tissues of vertebrates: essentially tricalcium phosphate and calcium carbonate [McConnel 1973].

The first structural identifications of biominerals using X-ray diffraction were obtained by de Jong in 1926 [de Jong 1926]. He established that calcium phosphate biominerals of vertebrates corresponded to an apatite structure and since that time bone mineral has been frequently identified as hydroxyapatite (HA):

$Ca_{10} (PO_4)_6 (OH)_2$

which was later considered to crystallize in the hexagonal system (space group $P6_3/m$), see on figure 1, and then shown to be monoclinic, at room temperature, when stoichiometric [Posner 1958, Elliott 1973]. The XRD data revealed very broad bands indicating low cristallinity. However, the identification of a poorly crystalline apatite structure did not explain the existence of significant amounts of carbonate in all mineralized biological tissues. Thus it was considered that bone mineral was a mixture of poorly crystalline apatite and amorphous calcium carbonate or that carbonate ions were at least in part adsorbed on the apatite crystal surface [Elliott 1994].

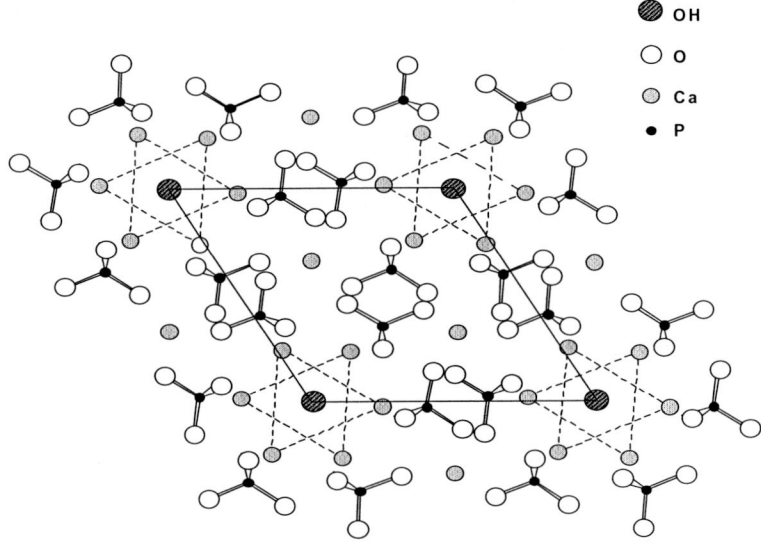

Figure 1. Projection on the (001) plane of hydroxyapatite structure. The two Ca^{2+} triangles lining the "tunnels" of the structure are located at z ¼ and ¾. OH^- ions are slightly under or above the triangles [Cazalbou 2004] - Reproduced by permission of The Royal Society of Chemistry.

It was only in the sixties that studies on synthetic carbonated apatites established that all carbonate ions could in fact be located within the apatite structure [Legeros 1968, Labarthe 1973]. But this has been the object of much controversy. Detailed studies indicated that carbonate ions could be located in the two anionic sites of the apatite structure: in PO_4^{3-} sites (type B carbonated apatite) and OH^- sites (type A carbonated apatite). Bone apatites were believed to correspond essentially to type B carbonated apatite whereas enamel contained both type A and B carbonate. The fraction of type A carbonate in dental enamel was evaluated at about 10% of the total carbonate content using infrared spectroscopy [Elliott 1985]. The use of carbonated apatite as a model for biological calcifications of hard tissues of vertebrates is nowadays accepted. The variability of the composition of apatite minerals and their mode of formation however needed further investigation.

From a kinetic point of view the direct formation of apatite crystals in calcified tissues was considered as unlikely, regarding the slow growth rate of apatite crystals and the existence in body fluids of crystal growth inhibitors such as magnesium and carbonate ions [Campbell 1991]. The improvement of crystallinity of bone apatites upon aging, revived for a while the theory of an amorphous phase considered as a necessary precursor in the formation of biological apatites [Termine 1966]. This hypothesis based on studies dealing with the formation and stability of amorphous calcium phosphates and its progressive conversion into apatite, compared to bone mineral evolution upon aging, was thus considered as quite consistent for a decade. In the eighties however it was demonstrated that bone mineral was composed of apatite nanocrystals with no or a non-detectable amorphous phase [Grynpas 1984]. The peculiar shape of bone crystals (platelets) showing non-equivalent a and b directions perpendicular to the c-axis of the hexagonal structure (figure 1), however led to the hypothesis that apatite formation involved another precursor phase, very close to apatite and indiscernible considering the very poor crystallization state of the mineral: triclinic OctaCalcium Phosphate (OCP) [Brown 1987]:

$$Ca_8 (PO_4)_4 (HPO_4)_2, 5H_2O$$

The OCP structure has been shown to consist in the association of an apatite-like layer and a hydrated layer [Mathew 1988]. Precipitating as platelet-shaped crystals, this phase exhibits a high crystal growth rate and hydrolyzes readily in aqueous media into apatite, forming interlayered compounds with hydroxyapatite and preserving the original platelet shape of the crystals. This model does not however explain all the variability of apatite compositions and particularly the very similar role played by carbonate and HPO_4^{2-} ions in bone mineral and synthetic analogues [Neuman 1956].

The chemical composition of biological apatite has been the object of several approximations frequently based on the composition of model minerals or synthetic analogues. A general chemical formula proposed by Winand for HPO_4^{2-}-containing apatite was [Winand 1961]:

$$Ca_{10-x} (PO_4)_{6-x} (HPO_4)_x (OH)_{2-x} \text{ with } 0 \leq x \leq 2$$

and by Labarthe et al. for carbonate-containing apatites [Labarthe 1973]:

$$Ca_{10-x}(PO_4)_{6-x}(CO_3)_x(OH)_{2-x} \text{ with } 0 \leq x \leq 2.$$

These formula establish a similar behavior for bivalent ion substitution of trivalent phosphates: the creation of a cationic vacancy and an anionic vacancy in monovalent sites. These chemical formulas are consistent with the limit composition observed (x=2) and the decrease of the OH⁻ content when the amount of carbonate and/or HPO_4^{2-} in the apatite increases. Other chemical formulas have been proposed. The most general one [Rey 2006]:

$$Ca_{10-x+u}(PO_4)_{6-x-y}(HPO_4^{2-} \text{ or } CO_3^{2-})_{x+y}(OH)_{2-x+2u-y} \text{ with } 0 \leq x \leq 2 \text{ and } 0 \leq 2u+y \leq x$$

is however of little relevance for biological apatites which are best approximated by the simple combination of the two previous formulas taking into account the possible existence of type A carbonates:

$$Ca_{10-x}(PO_4)_{6-x}(HPO_4 \text{ or } CO_3)_x(OH \text{ or } \tfrac{1}{2}CO_3)_{2-x} \text{ with } 0 \leq x \leq 2$$

The compilation of different cortical bone analyses suggests a relatively homogeneous composition [Legros 1987]:

$$Ca_{8.3}(PO_4)_{4.3}(HPO_4 \text{ or } CO_3)_{1.7}(OH \text{ or } \tfrac{1}{2}CO_3)_{0.3}$$

characterized by a very high vacancy content close to the maximum (x=1.7). The carbonate content varies with age: it is very low in embryonic bone, and can represent up to 80% of the bivalent ions in the bone mineral of old vertebrate animals. The OH⁻ content of bone is very low at any age and OH⁻ ions can barely be detected [Rey 1995, Pasteris 2004]. The composition of tooth enamel crystals reveals a radically different chemical composition:

$$Ca_{9.4}(PO_4)_{5.4}(HPO_4 \text{ or } CO_3)_{0.6}(OH \text{ or } \tfrac{1}{2}CO_3)_{1.4}$$

showing a much lower vacancy content, unveiling the unique adaptability of apatites to their biological functions [Cazalbou 2004].

Other minor substitutions are found in biological apatites involving for example trivalent cations (e.g. rare earth elements, actinides) or monovalent cations (especially Na^+) for Ca^{2+}, tetravalent ions replacing PO_4^{3-}, and bivalent ions replacing OH⁻. Several charge compensation mechanisms have been proposed. Although the ability of the apatite structure to fix many elements has several consequences regarding intoxications with mineral ions and diseases, such possibilities seem to have a minor influence on the chemical formula of apatites in calcified tissues due to the low amounts of these foreign ions [Iyengar 1999].

The present data underline the strong heterogeneity of bone mineral and apatitic biomineralizations. The global compositions do not reflect strong local variations between osteons and within osteons, and probably between the crystals themselves [Paschalis 1996]. In addition several properties of bone mineral such as ion exchange suggest the existence of surface modifications and possibly surface compositions different from the bulk at the level of a nanocrystal. The heterogeneities of the mineral are among its chief characteristics, mainly related to bone remodeling processes and formation conditions. Parameters susceptible to evaluate these characteristics would be of great utility.

The replacement and healing of damaged hard tissues have always been a concern for human beings as shown by the examination of mummies. It is however only very recently that calcium phosphates have been used for bone substitution and repair [Jarcho 1979]. The first to be used were stoichiometric hydroxyapatite (HA) and β-tricalcium phosphate (β-TCP) which are stable CaP at high temperature and can be easily sintered into ceramics. They are still the major industrial CaP biomaterials. β-TCP was shown to be bioabsorbable and replaced by bone whereas HA constituted non-degradable materials. β-TCP is mainly used as a bioceramic whereas HA is also being processed for other biomaterials uses such as the coating of metallic prostheses where it was found to considerably improve bone repair as an "osteoconductive" material or composite ceramic-polymer materials showing strong mechanical analogies with bone tissues and excellent bone bonding abilities [de Groot 1987, Bonfield 1988]. Biphasic Calcium Phosphates (BCP), associating these two high-temperature CaP allow a controlled resorption rate and have been reported to offer superior biological properties [Daculsi 2003, Legeros 2002]. They are progressively replacing β-TCP ceramics in Europe. A new technological step was made with the development of CaP cements [Brown 1986]. These materials are able to set and harden in a living body and most can be injected. Despite their poor mechanical properties they offer a number of advantages and are increasingly used for several applications. More recently biomimetic coatings involving low temperature nanocrystalline CaP have been proposed - some have been claimed to exhibit osteoinductive properties [Habibovic 2006].

Chapter 3

SYNTHESIS OF NANOCRYSTALLINE APATITES

As mentioned above, synthetic nanocrystalline apatites are of undeniable interest in the preparation of apatite-based bioceramics for bone substitution, repair or augmentation applications. However, synthetic apatites are also prepared and studied to better understand the formation of biological apatites and some of their properties [Legeros 1994]. Since the eighties, several synthetic routes have emerged and in the near future they could challenge the high-energy conventional processes involving high temperatures. Among the major advantages of these emerging unconventional processes are the use of low temperature (from room temperature (RT) to about 400°C), the flexibility and range of chemical compositions, and the physical, chemical and biological properties of these CaP.

Synthetic apatites can be prepared by several methods (precipitation under conditions of constant or changing composition, hydrolysis, solid/solid reaction at high temperature, hydrothermal methods) the type of which determines the amount and kind of substitution in the apatite. In this section we will focus on the synthesis of calcium-deficient apatites in solution systems.

Several processes (precipitation by double decomposition, sol-gel method, hydrolysis) involving various media (aqueous, hydro-alcoholic or organic solutions) leading to calcium phosphate apatites have been reported. Sol-gel processes still raise some problems: the long time needed for the preparation of the sol, and the presence of other calcium phosphate phases depending on aging time and temperature [Liu 2002]. Calcium-deficient or substituted apatites can also be prepared by hydrolysis of amorphous calcium phosphate, dicalcium phosphate dihydrate, octacalcium phosphate, or α and β tricalcium phosphate for example. Hydrolysis of these calcium phosphate phases to yield apatite can proceed through a dissolution-reprecipitation mechanism depending on the pH, the temperature and the presence of other ions. The latter can act as inhibitors (magnesium, pyrophosphate ions) or promotors (fluoride ions) of hydrolysis of dicalcium phosphate dihydrate (DCPD: $CaHPO_4$ $2H_2O$) and OCP for example [Legeros 1994].

Two major parameters determine the crystallinity and the calcium deficiency of the apatite obtained by precipitation methods: temperature (ambient temperature to 100°C) and pH (basic). When precipitated from solutions at temperatures between 80°C and 100°C, the higher the initial pH, the lower the calcium deficiency. Precipitation at temperatures under 80°C leads to less and less crystallized apatites, and the synthesis of poorly crystalline apatites analogous to bone mineral can be easily achieved at ambient temperature and physiological pH according to the method reported in the next sub-section.

Interestingly, precipitation using a hydro-alcoholic medium with a dielectric constant lower than that of water, provides control of hydrogenphosphate and carbonate ion content in apatite analogous to bone mineral [Zahidi 1985, Rodrigues 1998, Dabbarh 2000]. In addition, several studies reported the influence of the drying process and temperature on the composition, structure and degree of crystallinity of apatitic calcium phosphates [Dabbarh 2000, Lebugle 1986].

3.A. POORLY CRYSTALLINE APATITE (PCA) SYNTHESIS

The results presented in this chapter are related to poorly crystalline apatites synthesized at ambient temperature and physiological pH by double decomposition between a phosphate and carbonate solution (for example, 40g of $(NH_4)_2HPO_4$, 20g of $NaHCO_3$ and concentrated ammonia solution 1 ml in 500 ml of deionized water) and a calcium solution (fr example, 17.7g of $Ca(NO_3)_2$ $4H_2O$ in 250 ml of deionized water) as previously published [Rey 1989]. The calcium solution is rapidly poured into the phosphate and carbonate solution at room temperature (20°C) and stirred only for a few minutes. For investigations on freshly-precipitated nanocrystalline apatites, the precipitate is then very quickly filtered under vacuum and washed with deionized water (2 liters). Then the gel is freeze-dried and finally stored in a freezer to prevent further maturation of the PCA nanocrystals. This method leads to a carbonated poorly crystalline apatite analogous to bone mineral [Rey 1995]. Non-carbonated poorly crystalline apatite can be prepared by this method with a carbonate-free phosphate solution. Other recipes involving cationic and anionic solutions with slightly different Ca/P ratio and no concentrated ammonia solution also leads to poorly crystalline apatite. In all cases, the large excess of phosphate (and bicarbonate) ions in the solution provides pH buffering at pH = 7.4.

3.B. POORLY CRYSTALLINE APATITE (PCA) MATURATION

To study the physical-chemical properties of nanocrystalline apatites, the apatite can be left to mature after precipitation at room temperature in the mother solution without stirring and in a stoppered vial to minimize the release and uptake of CO_2 at physiological pH. This evolution in solution (maturation) is an important process that can help us understand the evolution of the composition, structure and properties of biological and synthetic biomimetic apatites after different aging times (corresponding to young and old bones for example).

After maturation during variable periods of time (from an hour to several months), the precipitates were filtered under vacuum and washed with deionized water. In the case of subsequent ion exchange experiments, part of the gel is then freeze-dried (reference sample) and the other part is used for ion exchange treatments.

The study of maturation properties of PCA is presented in section 6.A.

3.C. IONIC EXCHANGE (DIRECT AND INVERSE) ON POORLY CRYSTALLINE APATITE (PCA)

Two kinds of ion exchange experiments can be performed on PCA: anionic exchange (and the study of the reversibility of exchange between carbonate and hydrogenophosphate) ions, and cationic exchange (and the study of the reversibility of exchange between calcium and other cations such as strontium or magnesium ions as reported in this chapter). Also, such ion exchange experiments can be carried out on immature gels or on apatite samples matured for various durations.

The ion exchange is performed by exposing either the gel or the mature sample of nanocrystalline carbonated apatite (CA) to an "exchange" solution containing the target ion (HCO_3^- or HPO_4^{2-} for anionic hydrogenphosphate $\Leftrightarrow$ carbonate exchange, Mg^{2+} or Sr^{2+} for cationic exchange with Ca^{2+}) at varying concentrations (for example 1 M for 10 minutes). The starting salts used for the preparation of the "exchange" solution can be $NaHCO_3$, $(NH_4)_2HPO_4$, $Mg(NO_3)_2$ and $Sr(NO_3)_2$. Part of the "exchanged" samples are then filtered, washed with deionized water and freeze-dried. The other part is re-suspended for 10 minutes in the "inverse exchange" solution, containing the initial ion at a concentration of 1 M. In all cases, the samples are finally filtered, washed with deionized water and freeze-dried. In the text the notation "CA/X" represents a carbonated apatite exchanged with the ion X, and "CA/X/Y" represents the same sample after inverse exchange with the ion Y. For example, "CA/Sr" refers to a carbonated apatite for which part of the calcium ions has been exchanged with strontium ions, and "CA/Sr/Ca" refers to the same sample after inverse exchange of Sr^{2+} by Ca^{2+}.

The studies of ionic exchange properties of PCA are presented in section 6.B.

Chapter 4

CHARACTERIZATION OF APATITES

4.A. CHEMICAL ANALYSIS

Depending on the precipitation conditions, on the maturation time and/or on ion exchange treatments, the composition of calcium-deficient apatites can vary significantly. The determination of calcium, total phosphate and carbonate ions can be easily performed in different ways whereas the direct evaluation of hydrogenphosphate ions which are one of the most important markers of PCA and bone mineral crystals has not yet been possible (only indirect measurements are available). All the chemical analysis methods used for apatite characterization are based on the dissolution of apatite in acidic solution before the analysis (calcium and orthophosphate ions determination) or during the analysis (carbonate ions).

Calcium concentration can be determined by complexometry with EDTA and the phosphorus concentration by UV-visible spectrophotometry of the phospho-vanado-molybdenum complex. The Ca/P atomic ratio of apatites can be calculated from the result of these two analyses. The relative uncertainty on calcium and phosphorus concentrations has been evaluated at 0.5 %.

The titration of HPO_4^{2-} ions is more complex because the two inorganic orthophosphate ions, PO_4^{3-} and HPO_4^{2-}, encountered in apatites are in rapid equilibrium in solution. So the dissolution of apatite in acidic solution and the determination of total phosphorus content by colorimetry cannot distinguish HPO_4^{2-} and PO_4^{3-}. Therefore, the only way to determine the HPO_4^{2-} content is to condense the ions into pyrophosphates according to equation 1. Then the HPO_4^{2-} level is determined by chemical analysis using the Gee and Dietz method which is based on the formation of pyrophosphate in apatite-containing HPO_4^{2-} ions upon heating [Gee 1953]. During treatment of the apatite at 500°C for 3 hours (or

600°C for 20 min) the following reaction occurs leading to the formation of pyrophosphate ions [Gee 1955]:

$$2\ HPO_4^{2-} \rightarrow P_2O_7^{4-} + H_2O \qquad (eq.\ 1)$$

After thermal treatment, phosphorus atoms are titrated as orthophosphate at 460 nm by absorption spectrophotometry before (PO_4^{3-} only) and after (PO_4^{3-} and HPO_4^{2-}) acid hydrolysis of the P-O-P bond of pyrophosphate ions at 100°C for 1 hour. Thus, the level of condensed phosphate (therefore that of HPO_4^{2-}) is calculated from the difference of the results of these two analyses. The pyrophosphate content corresponding to the concentration of HPO_4^{2-} is determined within 0.5 %.

It shall be noted that this analysis method cannot be used to accurately quantify the HPO_4^{2-} content in bone and in synthetic carbonated apatites due to the presence of carbonate ions that can interfere with pyrophosphate ions and partially prevent $P_2O_7^{4-}$ formation according to equations 2 and 3 [Elliott 1994]:

$$2\ HPO_4^{2-} + CO_3^{2-} \rightarrow 2\ PO_4^{3-} + CO_2 + H_2O \qquad (eq.\ 2)$$
$$P_2O_7^{4-} + CO_3^{2-} \rightarrow 2\ PO_4^{3-} + CO_2 \qquad (eq.\ 3)$$

To open up a new alternative to the long and tedious analytical method of Gee and Dietz involving several steps and treatments, we recently set up a method to easily and rapidly evaluate the HPO_4^{2-} content in apatite, based on the mathematical decomposition of the v_4PO_4 band of the Fourier Transform InfraRed (FTIR) spectrum [Combes 2001]. Figure 2 shows the good correlation obtained between HPO_4^{2-} determined by chemical analysis] and by FTIR spectroscopy using Gee et al. and Combes et al. methods, respectively. However, the curve-fitting parameters (decomposition of the v_4PO_4 band, see on figure 5) need to be refined extensively to obtain a better correlation between these two methods, and thus apply this tool to rapid investigations on biological apatites.

The carbonate content of apatites was determined using a CO_2 coulometer (UIC Inc., USA) that measures the CO_2 released during sample dissolution in acidic conditions ($HClO_4$, 2M) and in a closed system. The CO_2 released is transferred into a photometric cell in a non-aqueous medium and titrated through an acid-base reaction [Huffman 1977].

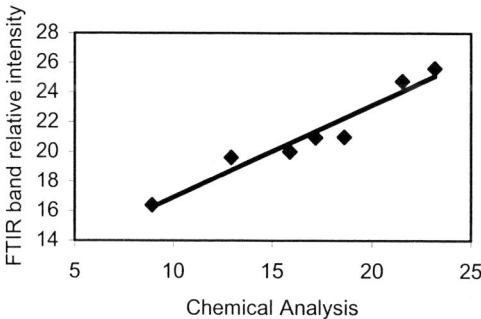

Figure 2. Correlation between HPO_4^{2-} content in apatite determined by chemical analysis and by FTIR spectroscopy [Combes 2001].

The determination of the amount of other ions taken up in PCA after ion exchange (strontium, magnesium ions for example) was performed using atomic absorption spectroscopy.

From the determination of the concentration of calcium, phosphate, carbonate and foreign ions if present (Mg^{2+}, Sr^{2+}), we calculated atomic ratios such as Ca/P, Ca/(P+C), or C/P in order to follow the evolution of the chemical composition of synthetic and biological PCA during maturation and/or after ion exchange processes (direct or inverse).

4.B. DIFFRACTION TECHNIQUES

The vast majority of the diffraction studies dealing with nanocrystalline apatites and reported in the literature is based on X-ray diffraction and only few electron diffraction data are available. This could be partly explained by the strong tendency for apatite nanocrystals to agglomerate leading to broad diffuse rings [Suvorova 1999]. Also, the way that electron beams affect these rather unstable compounds has not yet been established. To our knowledge, no neutron diffraction work has been dedicated so far to such poorly crystallized compounds.

The X-ray diffraction technique applied to the study of nanocrystalline apatite specimens is often primarily used to determine their apatitic phase purity. Although the presence of secondary crystalline phases such as pyrophosphates or whitlockite can generally be distinguished from the sharpness of their characteristic XRD patterns (within the detection limit of the equipment), the detection of other phases such as octacalcium phosphate (OCP), whose diffraction pattern is rather similar to that of hydroxyapatite, or amorphous calcium phosphate (ACP), generally requires special attention. The occurrence of a sharp low-angle diffraction peak around d=18 Å can however betray the presence of OCP for concentrations above the detection limit. A background halo in the range 27-40° (λ_{Co} = 1.78892 Å), and to a lesser extent 50-60°, is evidence of the presence of ACP. A quantitative comparison with the XRD patterns obtained for mixtures of known amounts of apatite and ACP can then be used for an estimate of the ACP amount present in the specimen. Pattern fitting methods can also be used for this evaluation. This was done for example by Rogers et al. who followed the amounts of ACP and nanocrystalline apatite present in coatings formed after immersion of a titanium substrate in simulated body fluid [Rogers 2005].

Beside phase purity evaluation, X-ray diffraction studies related to nanocrystalline apatites have mostly been dedicated to the evaluation of average crystal dimensions based on line broadening analysis [Arsenault 1988, Bonar 1983, Burnell 1980, Fisher 1987]. General findings indicate that the platelet-like apatite nanocrystals exhibit an average length along the c-axis in the range 200-400 Å for a thickness of about 20-80 Å. The width of a diffraction peak is indeed dependent on the size of the crystallites constituting the sample, following a 1/cosθ mathematical law (where θ is the diffraction angle) such as the Scherrer formula [Scherrer, 1918]:

$$L_{hkl} = \frac{K\lambda}{\beta_{size} \cos\theta_{hkl}} \quad \text{(eq. 4)}$$

where L_{hkl} is the average crystallite size perpendicular to the plane (hkl), λ is the X-ray wavelength, K is a constant close to unity dependent on the particle shape, and β_{size} is the line broadening due to the size effect and θ_{hkl} is the diffraction angle corresponding to the (hkl) plane.

However, two other factors also contribute to the overall line broadening: the existence of strain within the sample, giving a broadening effect following a tgθ law, and the instrument-related broadening effect. The latter can generally be evaluated from the XRD pattern of a well-crystallized reference sample such as stoichiometric hydroxyapatite (HA), for which size and strain broadening effects are considered negligible. The line broadening due to the sample itself (β_{sample}) can then be reached from the peak width observed after elimination of this instrumental contribution (β_{instr}). However, this process depends on the geometrical shape of the peak. Gaussian, Lorentzian (Cauchy), or a convolution of the two, are mathematical functions generally used for fitting X-ray diffraction patterns. For example, if the assumption is made that diffraction peaks can be satisfactorily fitted to a Gaussian function, then the line broadening due to the sample itself is given by equation 5:

$$\beta_{sample} = \beta_{size} + \beta_{strain} = \sqrt{\beta^2_{obs} - \beta^2_{instr}} \qquad \text{(eq. 5)}$$

where β_{obs} is the overall peak width observed. In this context, calculating average crystallite sizes from the sole application of the Scherrer formula requires the assumption that strain effects can be neglected, and therefore only leads to approximate values. This was emphasized in particular by Danilchenko et al. who investigated lattice strain and crystallite size in the direction of preferred orientations along the c-axis of the hexagonal unit cell in mature and well-mineralized cortical bone (femur) from a large animal (cow) [Danilchenko 2002]. These authors underlined the importance of considering lattice strain in peak profile analyses for a reliable estimation of the average crystallite size and proposed a method based on a threefold convolution of X-ray diffraction lines. They also proposed that crystallite size and strain parameters be considered as criteria for evaluating substructure variability among bioapatites. In the case of carbonated apatites synthesized at different temperatures, Baig et al. extracted the corresponding crystallite size and microstrain parameters from Rietveld analysis of XRD data, and correlated them to the evolution of the apatite metastable equilibrium solubility [Baig 1996, Baig 1999]. These authors concluded that microstrain rather than crystallite size was the dominant factor governing the solubility of such phases. The term microstrain may well in fact hide strong heterogeneities of apatite crystals composition. This phenomenon is not generally taken into account, however it is true for bone and can be related to the remodeling process. The heterogeneity in the composition of crystals induces a variation of their unit-cell dimensions and results in broadening of diffraction peaks related to the existence of unresolved superimposed peaks which can be misinterpreted as a decrease of crystal size and/or an increase of strain. Composition heterogeneity and the existence of variable solid-solutions also seem to occur for synthetic nanocrystals. The ability of apatite to give solid solutions and their surface equilibration necessities has led to the idea that even at the level of a crystal there could be some heterogeneity.

Theoretically, XRD peak analysis should be carried out by considering the integrated width based on the entire peak profile. In practice, the full-width at half-maximum (FWHM) of the considered peak is frequently used for the determination of β_{obs}. Also, whereas analysis of the whole pattern by Rietveld-like calculations leads to more accurate information, routine XRD studies from the literature usually only focus on selected diffraction lines. Due to the platelet shape of apatite nanocrystals and their elongation along the c-axis, rough estimates of crystallite size are often based on the two diffraction lines (002) and (310), see on figure 3. While the former leads to information on the length of the platelets, the latter gives an average value of their width/thickness. It must however be noted that for very immature apatites, lines (310) and (212) from the apatite structure tend to overlap due to the strong broadening size effect, which leads to a rather difficult or imprecise determination of the (310) peak FWHM. Eichert et al. reported estimated crystallite sizes drawn from the analysis of peaks (002) and (310) in the case of some biological and synthetic apatites (see table 1), and the results confirm the nanosize of these specimens [Eichert 2001]. Another example is given by Haris Parvez et al. who followed (002) and (310) line broadening for nanocrystalline apatites involved in composites containing increasing amounts of polyaspartate [Haris Parvez 2004].

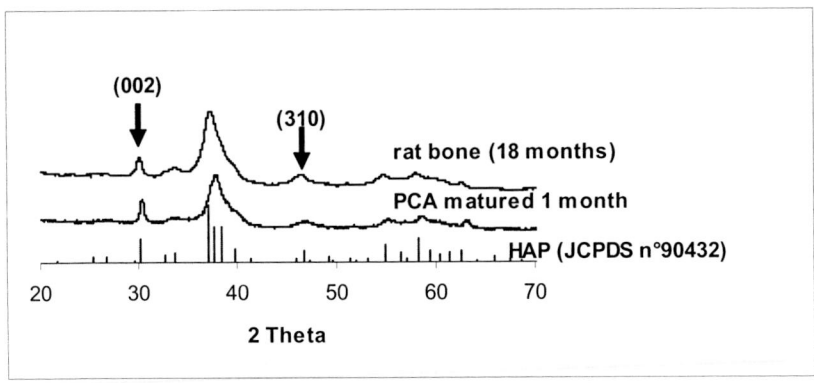

Figure 3. XRD patterns of hydroxyapatite (JCPDS reference), synthetic nanocrystalline apatite matured one month and rat bone. (λ_{Co} = 1.78892 Å).

The estimation of crystal size can be advantageously used to follow the maturation state (aging in solution) of a given apatite sample. The evolution of the XRD pattern toward better resolution is indicative of the increase in crystallinity of the sample during aging in solution (data not presented). However, the pattern obtained for the apatite sample matured for 1 month still shows broad peaks, betraying the rather limited increase of crystallite size over this period (see on figure 3). Basic profile analysis applying Scherrer's formula to lines (002) and (310), and ignoring strain effects, leads to L(002) values between 145 and 280 Å, and L(310) between 50 and 70 Å, for a maturation time of apatite ranging from 0 to 3 months [Eichert 2001]. The effect of temperature on nanocrystalline apatite crystallinity and crystal size can also be advantageously observed through XRD analyses. An example is given by Shirkhanzadeh et al. who studied nanocrystalline apatite coatings prepared by electrocrystallization [Shirkhanzadeh 1994]. This author reported an increase of the crystal size from 35 nm to approximately 100 nm when heating the coatings from the synthesis temperature (65°C) to 425°C.

Table 1. Estimated crystallite size for natural and synthetic apatites [Eichert 2001]

Sample	L(002) ± 3 Å (length)	L(310) ± 3 Å (width/thickness)
Chicken bone	207	66
Rabbit bone	190	Not evaluated
Adult human cortical bone	213	68
Synthetic apatite matured for 3 months	282	72

Nanocrystalline apatites are capable of accommodating high amounts of vacancies, leading to structural modifications of the regular apatitic lattice. The presence of structural defects can then be investigated by XRD analysis. For example, Wilson et al. performed Rietveld refinements of X-ray powder diffraction data in the case of calcium-deficient apatites, and reported the preferential loss of calcium from Ca(2) crystallographic sites rather than Ca(1) sites [Wilson 2005].

The determination of the hexagonal unit cell parameters "a" and "c" for a given specimen can give valuable information on its closeness to the regular apatitic lattice, by comparison with a reference sample like stoichiometric hydroxyapatite. XRD data are then particularly useful for following the evolution of the specimen under varying experimental conditions. For example, Panda et al. reported the increase of unit cell parameter "a" from 9.347 Å to 9.407 Å, thus approaching the reference value of 9.418 Å measured for stoichiometric HA when nanocrystalline hydroxyapatite samples initially synthesized at 80 °C by precipitation from a hydroxide gel were progressively heated up to 800°C [Panda 2003, Sudarsanan 1969]. In contrast, these authors did not observe variations of the "c" parameter in these conditions, and reported a value close to 6.862 Å to be compared to 6.881 Å for stoichiometric HA.

Variations of the lattice parameters can also be used to follow ion substitutions within the apatite structure. For example, Thian et al. recently reported in the case of a nanocrystalline hydroxyapatite coating on a titanium substrate that the substitution of Si^{4+} ions into the apatite structure resulted in an increase of both unit cell parameters "a" and "c" [Thian 2006]. These authors explained such trends by the greater ionic radius of Si^{4+} compared to P^{5+}. However, in this case, the discussion should also include composition variations related to the charge compensation mechanism. Despite this silicon substitution, no variation of XRD relative intensities was detected, although such effects are usually expected after lattice ion substitutions. This was justified here by the closeness of P and Si in the periodic table and the relatively low levels of silicon in substitution (up to 5 wt.%).

Finally, X-ray diffraction can be used for evaluating the Ca/P molar ratio of nanocrystalline apatites, and more globally of calcium-deficient apatites. The method is based on the evaluation of the amount of β-TCP formed upon heating the samples at temperatures higher than 700 °C, by comparison with XRD patterns obtained for known mixtures of β-TCP and apatite [Ishikawa 1993].

4.C. SPECTROSCOPIC METHODS

The most recent and significant progresses in the characterization of apatite nanocrystals were obtained by spectroscopic methods, especially Fourier Transform InfraRed (FTIR), Raman and Solid State Nuclear Magnetic Resonance (NMR) spectroscopies.

FTIR and Raman Spectroscopies

Vibrational spectroscopies (FTIR and Raman) can deliver information on the chemical environments of phosphate, carbonate, water molecules and hydroxide ions. The theoretical vibrational modes of phosphate groups in stoichiometric apatites are shown in table 2.

Detailed studies have distinguished and identified most of these bands in well-crystallized apatites [Leung 1990, Penel 1998]. In highly substituted non-stoichiometric apatites however the assignments are much less precise. Distortions of the ionic environments introduce band broadening limiting the band resolution and partly disrupt the vibrational correlation related to factor-group theory.

Nanocrystalline apatite spectrum is showed in figure 4 and the phosphate band positions are reported in table 3 (comparison with stoichiometric HA).

Table 2. Theoretical internal vibrational modes of PO_4^{3-} ion in stoichiometric HA (R: Raman activity, IR: Infrared activity).

v_1	v_2	v_3, v_4
Hexagonal, Space Group: $P6_3/m$		
Site symmetry (C_s)		
1 A' (IR,R)	1A' (IR,R), 1A"(IR,R)	1A' (IR,R), 2A"(IR,R)
Factor group theory (C_{6h})		
A_g(R), E_{2g}(R), B_u, E_{1u}(IR),	$1A_g$(R), $1E_{2g}$(R), $1B_u$, $1E_{1u}$(IR), $1B_g$, $1E_{1g}$(R), $1A_u$ (IR), $1E_{2u}$	$2A_g$(R), $2E_{2g}$(R), $2B_u$, $2E_{1u}$(IR), $1B_g$, $1E_{1g}$(R), $1A_u$ (IR), $1E_{2u}$

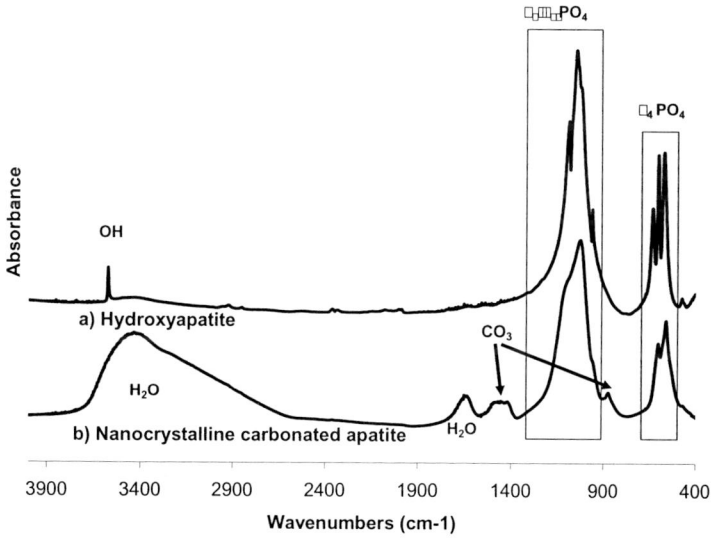

Figure 4. FTIR spectrum of a) well-crystallized hydroxyapatite and b) poorly crystalline carbonated apatite.

When HPO_4^{2-} ions are present, specific bands relative to this ion are observed.

In most nanocrystalline apatite samples and especially in biological apatites, however, additional bands are observed which do not appear in well-crystallized apatites and which have been designated as "non-apatitic environments" of the mineral ions (figure 5) [Rey 1989, Rey 1990]. These "non-apatitic" phosphate environments have been shown to appear more clearly in the v_4 PO_4 domain.

Table 3. IR and Raman bands observed for stoichiometric HA and nanocrystalline apatite

Domain, Assignments	Stoichiometric Hydroxyapatite		Nanocrystalline apatite	
	IR (cm^{-1})	Raman (cm^{-1})	IR (cm^{-1})	Raman (cm^{-1})
ν_2 PO$_4$		433		432
	464	448	469	452
	474			
HPO$_4$ non-apatitic			533	
HPO$_4$ apatitic			551	
ν_4 PO$_4$	567	580	562	584
	572	591	575	590
	602	607	603	611
		614		
PO$_4$ non-apatitic			617	
ν_L OH	633			
P—OH of HPO$_4$			870	873
non-apatitic			866	
ν_2 CO$_3$ type B			871	
type A			880	
ν_1 PO$_4$	964	964	962	961
ν_3 PO$_4$			1006	1005
			1020	
	1026	1029	1031	1032
	1034	1034		
	1044	1041	1044	1044
		1057	1059	
	1063	1064		
	1089	1077	1072	1071
			1091	
			1104	
HPO$_4$			1144	
ν_1 CO$_3$ type B				1071
ν_1 CO$_3$ type A				1103
B + *non-ap*			1420	
ν_3 CO$_3$ A + B			1460-1470	
non-apat			1500	
A			1540	

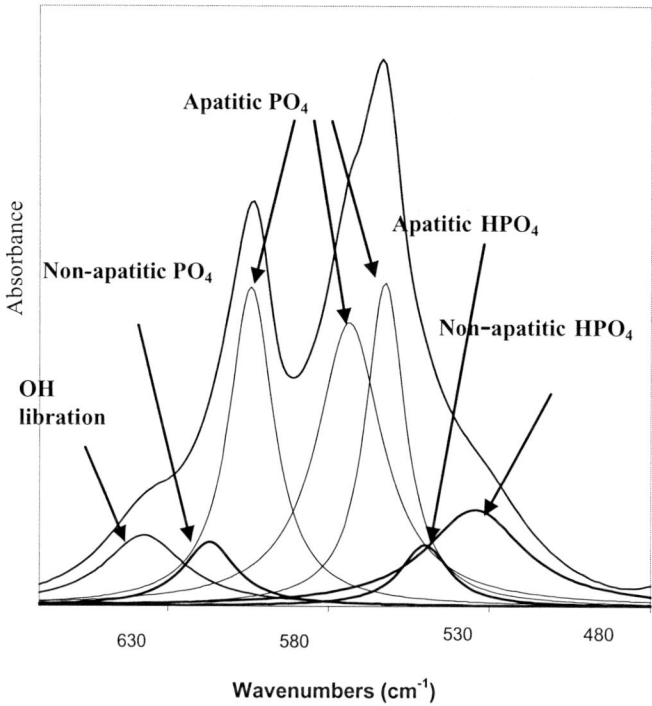

Figure 5. Example of $\nu_4 PO_4$ FTIR band decomposition (curve-fitting) using Grams/32 software (Galactic Industries Corp.).

Two distinct shoulders corresponding to measurable bands after curve-fitting have been respectively assigned to non-apatitic PO_4^{3-} groups and non-apatitic HPO_4^{2-} groups. The last assignment was confirmed using chemical analysis (see section 3.A) [Combes 2001]. The formation of non-apatitic environments has been shown to be related to the synthesis of apatite nanocrystals at physiological pH. It corresponds to exchangeable surface ions (see in the sections 4 and 5).

Carbonate ions in pure type A and type B environments exhibit specific FTIR and Raman bands (table 3). However, as in the case of phosphate groups, additional vibrational bands are found in all biological apatites and apatite nanocrystals synthesized at physiological pH, corresponding to non-apatitic environments of the carbonate ions (figure 6 and table 3). A characteristic band is clearly seen in the $\nu_2 CO_3$ IR domain and can be used for the quantitative determination of the amount of "non-apatitic" carbonate environments. The latter have been shown, using ion exchange experiments, to share the same surface domain as the "non-apatitic" HPO_4^{2-} environments (see in the sections 4 and 5).

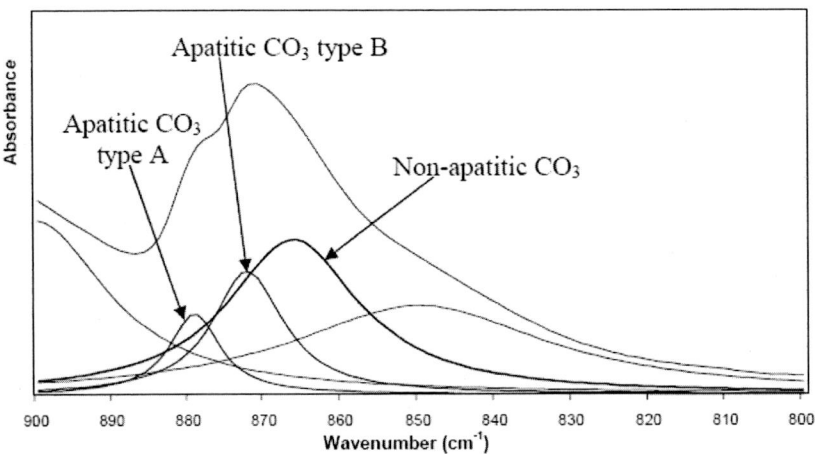

Figure 6. Example of v_2CO_3 FTIR band decomposition (curve-fitting) using Grams/32 software (Galactic Industries Corp.).

Most characterization procedures of biomaterials have been developped for dry samples. However, biomaterials function in aqueous media, and, especially for nanocrystals, the hydrated surface might considerably differ from dry surfaces. Thus, experiments have been recently carried out on wet samples to determine possible alterations of their spectroscopic characteristics.

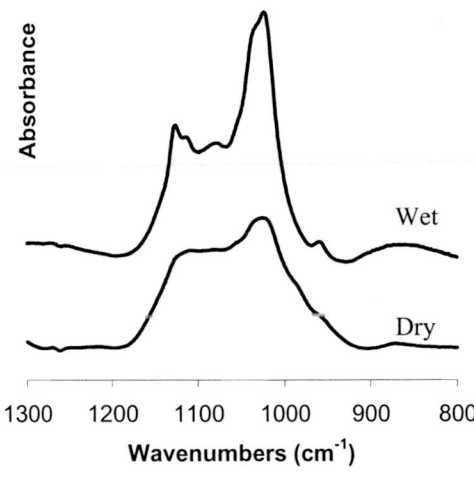

Figure 7. Effect on drying on the ionic environment as shown by FTIR spectroscopy in the $v_3PO_4^{3-}$ domain. Reprinted from [Rey 2006] with permission from Elsevier.

The data obtained show a considerable change (see on figure 7). The wet, freshly precipitated apatite nanocrystals exhibit very thin bands testifying to the existence of a structured hydrated layer whose characteristics seem close to those of OCP, although several differences have been noticed [Eichert 2004]. Upon drying, this fine structure is lost and considerable band broadening occurs leading to the features already described as "non-apatitic environments".

The structuration of the hydrated layer seems very sensitive to the ion content of the nanocrystal surface and it has been shown to be reversibly modified during ion exchange reactions (see section 5B). Some of the characteristics of the wet crystals spectra are summarized in table 4. These characteristic bands decrease progressively (but they never totally disappear) during the maturation of the wet crystals confirming the slow growth of apatite domains at the expense of the surface hydrated layer [Eichert 2001, Eichert 2004].

Table 4. Positions et assignments of FTIR bands for wet precipitated PCA (w: weak, vw: very weak; sh: shoulder).

Assignments	Band positions (cm^{-1})
HPO_4 (OH in plane bend)	1195 w,sh
Stretching v_3 HPO_4	1137 w,sh
	1127
	1110
Stretching $v_3 HPO_4$, v_3 PO_4	1075 w,sh
Stretching v_3 PO_4	1055 w,sh
Stretching v_3 PO_4	1035
Stretching v_3 PO_4	1022
Stretching $v_1 HPO_4$	1000 vw,sh
Stretching $v_1 PO_4$	961
Stretching HPO_4 (P-OH)	860

Solid State NMR

For over twenty years solid state NMR has been widely used to investigate the chemical structure (bulk and surface) and structural properties of calcium phosphates and related biomaterials. In particular, this technique provides detailed information on chemical and structural environments of phosphate, carbonate and hydroxyl groups, by observing single and/or double resonance of different nuclei

(^{1}H, ^{31}P and ^{13}C) [Rothewel 1980, Tropp 1983, Aue 1984, Yesinovski 1987, Belton 1988, Miquel 1990, Beshah 1990, Pan 1995, Cho 1996, Sfihi 2002, Wilson 2005, Jarlbring 2006, Jäger, 2006]. In a stoichiometric apatite and related materials, the ^{31}P chemical shift of all the phosphate groups has been reviewed [Yesinovski 1998]. In stoichiometric apatite, all PO_4^{3-} groups have the same chemical environment and therefore only a single line is observed at 2.3-2.9 ppm with respect to 85 % H_3PO_4 [Rothewel 1980, Tropp 1983, Aue 1984, Belton 1988, Miquel 1990, Sfihi 2002, Wilson 2005, Jarlbring 2006, Jäger, 2006]. However, in some cases, in particular when the samples are not well crystallized, this line is separated into two components corresponding to PO_4^{3-} groups in crystalline and amorphous phases, by combining relaxation and double resonance experiments [Isobe 2002]. In bone nanocrystals, the study of chemical shift anisotropy suggests P environments different from those of apatites [Roufosse 1984, Roberts 1992, Wu 1994]. In addition, in apatite nanocrystals an inhomogeneous and broad ^{31}P peak of the main phosphate is generally observed [Wu 1994, Jarlbring 2006]. This peak, related to HPO_4^{2-}, can be isolated using 2-D ^{1}H↔^{31}P solid state NMR [Wu 1994].

^{31}P solid-state NMR spectra of wet nanocrystals are very different from those of dry crystals. Indeed, they exhibit different peaks located in the 0 - 6 ppm range assigned to different phosphate species [Eichert 2004]. These specificities are consistent with the FTIR observations, but the analogy with OCP, as could have been expected, is not attested. The HPO_4^{2-} peak seems the most affected suggesting that these species are essentially located in the hydrated layer (see section 4). The characteristics of the spectra suggest possible proton hopping between phosphate groups of the hydrated layer.

Because of the very low carbonate content, the studies of carbonate ions generally needed enrichment with ^{13}C for NMR observation in a reasonable time. Type A and Type B carbonated apatites show two distinct peaks located at 166 and 170 ppm with respect to tetramethylsilane (TMS), respectively [Beshah 1990]. In addition "non-apatitic" carbonate can exhibit a broad peak at ca. 168 ppm which has been shown to be associated with water molecules [Sfihi 2002].

^{1}H solid state NMR measurements have been performed especially to detect hydroxide ions. In hydroxyapatite, the OH$^-$ groups give a single ^{1}H NMR peak at approximately 0.2 ppm with respect to TMS [Yesinowski 1987]. This peak has been detected at the same position in bone [Cho 2003]. However, the indirect ^{1}H solid state NMR measurements used to detect the OH groups in bones did not allow an evaluation of their amount. Other ^{1}H peaks are related to water molecules again associated with bone apatite nanocrystals [Cho 2003]. Similar results were recently reported in synthetic nanocrystalline apatites [Cho 2003].

This peak, which appears in most calcium phosphate bioamaterials and which is centered at *ca.* 6 ppm, is relatively broad and probably corresponds to several labile environments. It seems dominated by water molecules from the surface layer (see the model presented section 4). Additional unresolved ^{1}H broad peaks of weak intensities assigned to HPO_4^{2-} ions can also be observed in the range 10 – 16 ppm both in bone mineral [Cho 2003] and in synthetic nanocrystalline apatites [Jäger 2006]. These peaks testify to the existence of several HPO_4^{2-} environments, but their study needs to be completed, to specify in particular their localisation.

X-Ray Absorption Spectroscopies (XANES and EXAFS)

As illustrated above, the environments of anions in nanocrystalline apatites can be characterized using different spectroscopic techniques. In contrast, the environments of calcium ions can only be observed through techniques involving core electron transitions which are rather difficult to carry out. Several sets of data have however confirmed the existence in non-stoichiometric nanocrystalline apatites of specific cationic environments that are absent in well crystallized stoichiometric apatites and assigned to "non-apatitic" environments of calcium ions [Eichert 2005]. The existence of these environments can be related to cation exchange properties of apatite nanocrystals.

Chapter 5

A MODEL FOR NANOCRYSTALLINE APATITES

Nanocrystalline apatites, whether biological or synthetic, exhibit an extended surface area due to the nanosize of their constitutive crystals. As for any other kind of nanomaterials, the surface-to-volume ratio is therefore high and all experimental results must then be considered as a combination of bulk and surface contributions. The importance of surface behavior is even more crucial for biomedical applications since numerous functions of the bone mineral involve at the interface between the surface of such apatite nanocrystals and the surrounding biological fluids.

In addition to the use of characterization techniques specifically adapted to surface investigations, experimental results obtained with non-surface-specific techniques can also be exploited and were indeed found to be of prime importance for the exploration of the nanocrystalline apatites surface structure. Section 3 reported results of our investigations using spectroscopic techniques (FTIR and solid state NMR) revealing the presence of non-apatitic chemical environments in the case of synthetic as well as biological nanocrystalline apatites.

We can summarize that according to the results of complementary investigations on nanocrystalline apatites using spectroscopic techniques and chemical analysis methods presented in the previous section: (1) apatite nanocrystals involve non-apatitic anionic and cationic chemical environments, (2) at least part of these environments (e.g. labile carbonates) are located on the surface of the nanocrystals and are in strong interaction with hydrated domains, (3) immature samples show IR band fine substructure that is altered upon drying without leading to long-range order modifications, and (4) this fine substructure shows striking similarities with the FTIR signature of OCP which is constituted by alternating "apatitic" and "hydrated" layers.

All these elements favor a model in which apatite nanocrystals are covered with a rather fragile but structured surface hydrated layer containing relatively mobile ions (mainly bivalent anions and cations: Ca^{2+}, HPO_4^{2-}, CO_3^{2-}) in "non-apatitic" sites. However the exact structure and composition of this hydrated layer are still under investigation. The existence of a hydrated layer with a composition different from that of bulk apatite domains can modify the description of apatite composition presented in section 1. A schematic representation of this "surface hydrated layer model" is given on figure 8.

It is interesting to remark that the two calcium phosphate phases with which spectroscopic similarities were found (DCPD and OCP) are hydrated phases and exhibit the lowest surface tensions among the examples of calcium orthophosphates studied in the literature. One could thus postulate that the formation of a hydrated layer on the surface of apatite nanocrystals could contribute to the reduction of the surface energy of the nanocrystals.

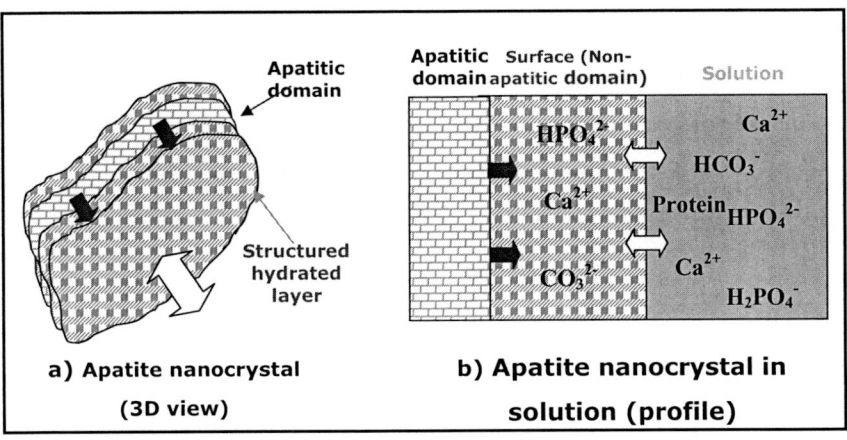

Figure 8. Schematic representation of the " surface hydrated layer model" for poorly crystalline apatite nanocrystals.

This layer might be close although not identical to that existing in octacalcium phosphate [Eichert 2004]. It differs from OCP by its ability to accomodate carbonate ions and other mineral ions like Mg^{2+}. It is thought to be responsible for most of the properties of apatites, especially high electrical conductivity probably resulting from a high superficial ionic mobility. The consideration of this type of surface state can help to understand and explain the behavior of biological apatites in participating in homeostasis due to the very high specific surface area of bone crystals and also in constituting an important ion reservoir with an availability that depends on the maturation state. The consequences of this particular mineral structure have not been completely explored, it is however plausible that bone mineral is also involved in some hormonal regulations or in chemical interactions with the organic matrix that could determine its mechanical properties. Nanocrystalline apatite properties are presented in detail in the next section.

Chapter 6

PHYSICO-CHEMICAL PROPERTIES OF NANOCRYSTALLINE APATITES

6.A. MATURATION

Poorly crystallized, non-stoichiometric apatites are not thermodynamically stable and show a clear tendency to evolve in solution toward stoichiometry and greater crystallinity [Cazalbou 2004]. As indicated by the arrows in figure 8a, upon aging in solution the bulk apatitic domains slowly extend at the expense of the hydrated domains at the surface of the nanocrystals and can irreversibly incorporate some of the ions of the hydrated layer. This evolution is referred to as "maturation" or "aging in solution". It can be followed by several physico-chemical measurements.

The overall chemical composition of the initial phase varies with maturation time. For example, the changes in composition observed for a carbonated nanocrystalline apatite as a function of maturation time in the synthesis solution at room temperature (RT) are reported in table 5.

In this case, the increase of the Ca/P molar ratio from 1.44 to 1.62 during the first steps of maturation is accompanied by the constancy of the Ca/(P+C) ratio and the increase of the C/P ratio. These trends indicate that over this time period the total number of vacancies remains unchanged for carbonated apatites while HPO_4^{2-} ions get replaced by CO_3^{2-} ions. These conclusions are confirmed by FTIR spectroscopy analysis (figure 9a and 9b), showing the progressive decrease of the non-apatitic HPO_4^{2-} content (and labile carbonates) and the concomitant increase of apatitic carbonate species (mostly B-type).

Table 5. Evolution of the chemical composition (atomic ratios) of a carbonated nanocrystalline apatite with maturation time (in synthesis solution, at RT).

Maturation time	Ca/P (± 0.02)	Ca/(P+C) (± 0.02)	C/P (± 0.02)
0	1.44	1.41	0.030
6 hours	1.51	1.42	0.060
1 day	1.51	1.40	0.089
3 days	1.54	1.39	0.145
10 days	1.62	1.40	0.162
4 months	1.62	1.38	0.200

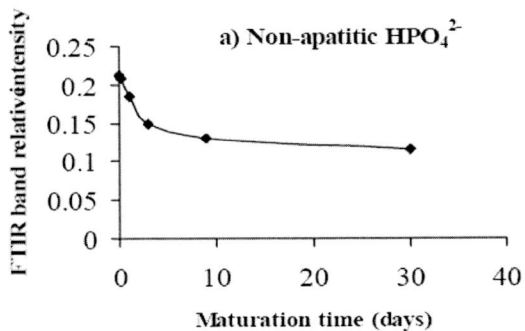

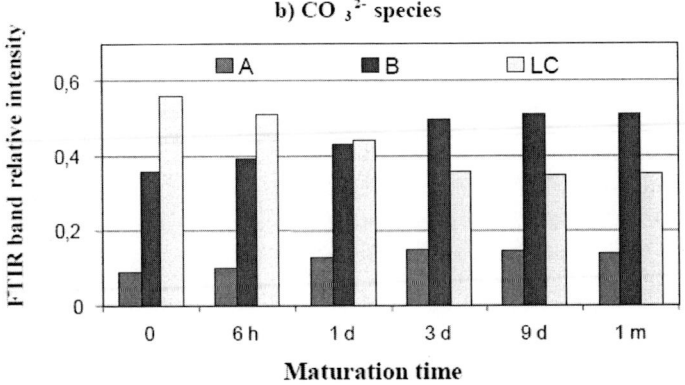

Figure 9. Evolution of a) labile HPO_4^{2-} and b) CO_3^{2-} (A: A-type carbonate, B: B-type carbonate, LC: labile carbonate species) contents with maturation time for a carbonated nanocrystalline apatite, as determined by FTIR spectroscopy.

Maturation is a physico-chemical process, initiated by a thermodynamical driving force, and leading to modifications of the phase composition and structure. As such, it should be distinguished from Ostwald's aging. The initial nanocrystalline phase, rich in (surface) non-apatitic environments and poor in carbonate evolves toward a more stable structure with growing apatitic domains, in contrast with the surface hydrated layer which tends to disappear. The structure evolves toward higher stability and therefore lower solubility. The chemical composition of synthetic apatites is thus controlled by the synthesis conditions.

In parallel to chemical composition changes upon maturation, other physico-chemical characteristics also evolve. In particular, an increase in crystallite size can be clearly seen (figure 10).

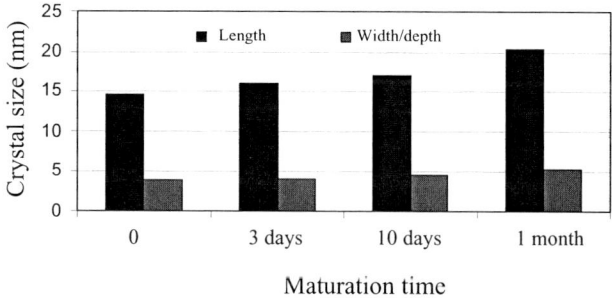

Figure 10. Variation of nanocrystalline apatite crystallite size with maturation time.

In this context, the maturation state of nanocrystalline apatites has to be considered as a major parameter in the preparation of nanocrystalline apatite-based biomaterials, as it controls important factors after implantation such as dissolution rate. The surface reactivity (see next section) is also modified by the apatite maturation.

In parallel to chemical composition changes upon maturation, other physico-chemical characteristics also evolve. In particular, an increase in crystallite size can be clearly seen (figure 10).

In this context, the maturation state of nanocrystalline apatites has to be considered as a major parameter in the preparation of nanocrystalline apatite-based biomaterials, as it controls important factors after implantation such as dissolution rate. The surface reactivity (see next section) is also modified by the apatite maturation.

6.B. Ion Exchanges on Nanocrystalline Apatites

Ion exchanges in bone mineral have been the object of numerous studies [Neuman 1956, Pak 1967, Neuman 1968, Johnson 1970, Fernandez Gavarron 1978, Neuman 1985]. Indeed, these processes have been suggested to be involved in the regulation of mineral ion concentrations in body fluids (homeostasis). In this view, the bone mineral can be considered as a "reservoir" for mineral ions, capable of entrapping or releasing them as required for concentration regulation. The above-cited studies concerned both cationic (calcium) and anionic (phosphate and carbonate) exchanges. Unsurprisingly, such ion exchange processes originate at the interface between apatite nanocrystals and the surrounding biological fluids.

The first studies of ion exchange phenomena on bone apatite were mostly dedicated to phosphate/carbonate substitutions, and these investigations were first initiated in view of explaining the presence of carbonate ions in bone. Since then, the incorporation during the mineralization process, of CO_3^{2-} ions into the apatitic lattice by substitution of either phosphate (B-type) or hydroxyl (A-type) ions has been identified and studied in detail (see section 3c).

Non-apatitic carbonate environments, also referred to as "labile carbonates", have been identified in immature samples (see figure 6 and table 3). Recent works have pointed out the possibility for nanocrystalline apatites to exchange such labile CO_3^{2-} ions with phosphate groups through ion exchange in a solution enriched in a phosphate salt. Hina et al. have investigated such exchanges, and found that soaking a nanocrystalline apatite sample in a carbonate-rich solution led to an increase of the carbonate content of the solid accompanied by a decrease of its non-apatitic HPO_4^{2-} content [Hina 1996]. Similarly, when the precipitate is immersed in a phosphate-rich solution, the amount of non-apatitic HPO_4^{2-} contained in the solid increased at the expense of the carbonate content. Interestingly, such non-apatitic carbonate/non-apatitic HPO_4^{2-} exchanges were found to be highly reversible (figure 11). This is an additional point in favor of surface exchange phenomena involving easily accessible sites, most probably within the nanocrystals superficial hydrated layer. It is important to note that the apatitic carbonate species are not altered by subsequent immersion in phosphate solution. This was for example shown by Hina et al. using $^{13}CO_3^{2-}$ ions [Hina 1996]. This emphasizes the distinction between the ions that are located in the surface layer and are therefore rather mobile, and those located in stable apatitic sites in the bulk (see figure 8 and section 4).

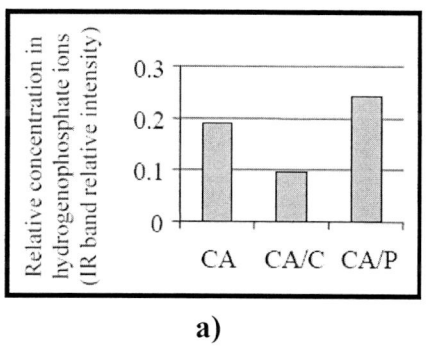 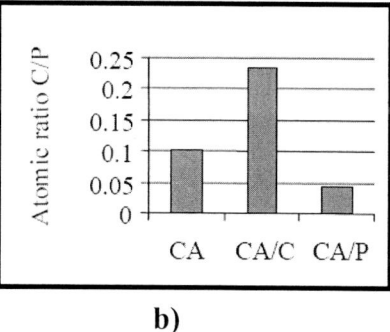

a) b)

Figure 11. Example of ionic exchange (carbonate-phosphate) experiments: a) Evolution of hydrogenphosphate ion concentration (IR data) b) Evolution of carbonate ion concentration (C/P ratio calculated from chemical analysis results). CA: initial carbonated apatite matured 24h, CA/C: CA immersed in ammonium carbonate solution (1M, 10 mn); CA/P : CA immersed in ammonium phosphate solution (1M, 10 mn) [Cazalbou 2004] - Reproduced by permission of The Royal Society of Chemistry.

The capacity to undergo such surface ion exchanges is however highly dependent on the maturation state of the apatite specimen. Indeed, the total amount of exchangeable carbonate ions (i.e. labile carbonates) was found to decrease with maturation time as indicated in figure 12. This phenomenon is attributed to the progressive disappearance, upon maturation, of the hydrated surface layer of nanocrystalline apatites which is thought to accommodate the labile carbonate species.

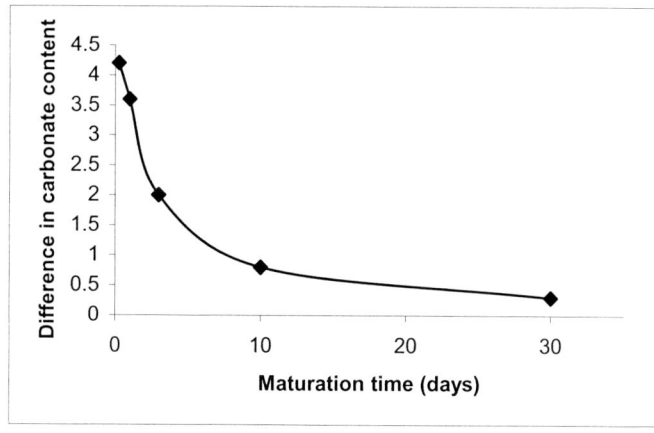

Figure 12. Decrease in exchangeability of carbonate ions versus maturation time.

Beside anionic exchanges, illustrated above by the carbonate/phosphate pair, cationic exchanges (substitution of calcium) also deserve special attention. In this regard, ions such as Mg^{2+} and Sr^{2+} are of particular interest due to their biological activity. Magnesium is one of the most abundant foreign elements in biological hard tissues and is thought to play a role on bone cell adhesion [Tsuboi 1994]. Strontium, although a minor constituent of bone, was shown to play an important role in osteoclast cell activity and could therefore be used in the treatment of some bone pathologies [Pors 2004]. Marie et al. showed for example that strontium had pharmacological effects, including antiresorptive and anabolic activity, leading to possible interest for the treatment of osteoporosis and other osteopenic pathologies [Marie 2001].

The works reported by Cazalbou et al. and Eichert et al. show that magnesium and strontium ions can be entrapped by nanocrystalline apatites by immersion in a solution containing the exchanging ion [Cazalbou 2000, Eichert 2001]. The amounts of Mg^{2+} and Sr^{2+} taken up were found to decrease with apatite maturation (figure 13). The reversibility of these exchanges was also investigated. Most of the magnesium taken up after ion exchange in a Mg-rich solution was fixed in a reversible way whatever the maturation stage of the apatite (see on figure 13a). In contrast, the amount of reversibly-fixed Sr^{2+} ions decreased noticeably with maturation in a Sr-containing solution (see on figure 13b). These results suggest that Mg^{2+} ions are mostly entrapped on surface sites whereas Sr^{2+} ions are progressively irreversibly incorporated (beyond 24 hours of maturation) into the apatitic domains of the maturating nanocrystals.

Recently the cationic exchange ability of non-carbonated and a carbonated nanocrystalline apatite, matured for one day, for magnesium and strontium ions was studied in depth. The Langmuir-like isotherms for magnesium ions are reported in figure 14, and indicate that the maximum amount of Mg exchanged is greater for the carbonated sample. These results were attributed to a greater proportion of hydrated layer on the carbonated apatite sample, due to the well-known apatite crystal growth inhibiting effect of carbonate ions. Similar results are obtained for strontium ions but Sr uptake is noticeably greater than that of Mg for both kinds of apatites (data not presented). The difference in behavior of Mg and Sr ions is still under consideration. Possible reasons include the occupancy of different cationic surface sites and differences in stability of the complexes $[Me(H_2O)_6]^{2+}$ (Me = Mg, Sr) in solution due to the larger size of Sr^{2+} ions.

The reverse exchanges were also investigated by soaking the pre-exchanged samples in a calcium-rich solution. In these conditions, ca. 85% of the Mg ions and 75-80% of Sr ions were released upon inverse exchange, indicating that on apatites matured one day most of the magnesium and strontium remained exchangeable, and therefore most probably located on surface sites within the nanocrystal hydrated layer rather than in apatitic domains of the bulk. It is also important to note that no secondary phase was observed by XRD analysis after such ion exchange experiments.

As a concluding remark on ion exchanges on nanocrystalline apatites, it is of interest to distinguish the two categories of exchange ions. A first type of ion can be incorporated into the surface hydrated layer of the nanocrystals and can then progressively replace apatitic ions in the bulk. This is the case of strontium. The second type of ion cannot massively substitute for bulk ions and they mostly remain located at the nanocrystal surface, like magnesium. These considerations, once generalized, have several implications especially for biological nanocrystalline apatites. As apatite maturation is an inevitable physical-chemical phenomenon leading to the loss of vital ion exchange properties, it is essential that bone mineral is renewed [Rey 1995]. However with aging, the bone remodeling cycle becomes slower, and therefore superficial properties are predictably degraded. This evolution explains the higher sensitivity of children to poisoning by some mineral ions and the preferential location of these contaminants in areas of fast bone remodeling (epiphysis) [Cazalbou 1999]. For example, intoxication with heavy metals is a well-known health issue especially problematic for young children and among such elements is lead (Pb^{2+} ions exhibiting an ion radius close to that of strontium). It is therefore likely that these ions could reach the surface of apatite nanocrystals through blood transport, and be entrapped in the surface sites. The higher sensitivity of children to lead intoxication could then be due to their less matured bone mineral, offering greater proportions of non-apatitic environments available for rapid ion exchanges. This stresses the importance of rapid intervention after such lead poisoning, in order to proceed with inverse exchange with calcium before the Pb^{2+} ions reach stable apatitic substitution sites. Indeed, nearly 50% of any strontium uptake is inaccessible to further ion exchange after maturation of 24 hours [Cazalbou 2000].

6.C. PROTEIN ADSORPTION ON NANOCRYSTALLINE APATITES

A good knowledge of the bone mineral physico-chemical characteristics is essential to understand its biological functions as well as to develop engineered biomaterials aimed at bone repair. However, this needs to be completed by the analysis of surface reactivity toward biological fluids. The previous section concerned mineral ion exchanges involving the surface hydrated layer of apatite nanocrystals. Other major components of body fluids are proteins. Since protein adsorption is known to precede cell adhesion to bone apatite, such adsorption processes also appear as critical steps and deserve special attention.

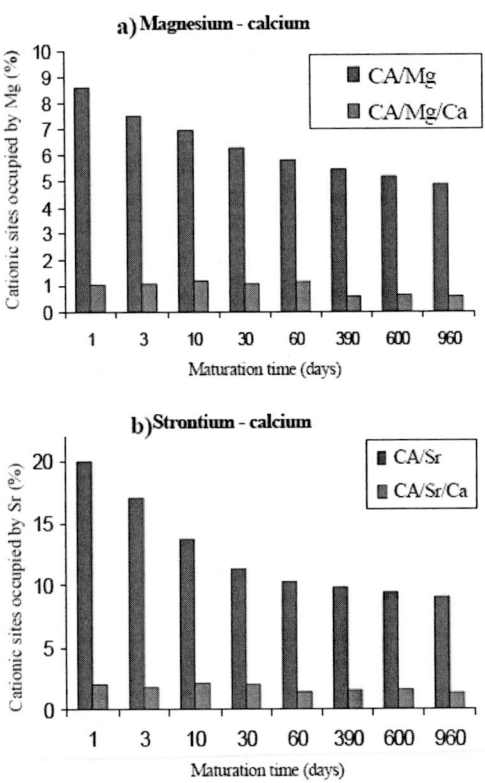

Figure 13. a) Mg and b) Sr uptake versus nanocrystalline apatite maturation time.

The first interest in protein adsorption arose from the observation that mineralized matrix proteins could initiate mineralization and regulate crystal growth. Several studies dealt with non-collagenous proteins such as albumin or phosphoproteins [Termine 1980, Glimcher 1990]. Attempts to classify these proteins as biomineralization promoters or inhibitors failed as other parameters like protein concentration and conformation were found to play a key role in such processes. Combes et al. reached similar conclusions during the study of the effect of bovine serum albumin (BSA) on the nucleation and growth of octacalcium phosphate (OCP) on type-I collagen. In this case, BSA was found to play alternately a promoter effect at low concentrations and an inhibitor effect at high concentrations (> 10 g.L^{-1}) [Combes 2002].

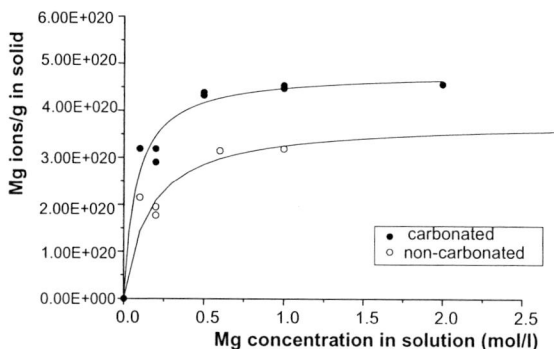

Figure 14. Magnesium uptake on 1 day-maturated nanocrystalline apatites (carbonated or non-carbonated) after direct exchange with Ca^{2+}.

In contrast to the large number of studies reported on protein adsorption on well-crystallized hydroxyapatite or other calcium phosphates, only few data are available for nanocrystalline apatites that constitute the most abundant mineral component of bone tissue. Ouizat et al. investigated the effects of BSA adsorption on such poorly crystallized apatites [Ouizat 1999]. In this work, the authors concluded that non-apatitic environments, in particular labile phosphate groups, were involved in the protein binding process, and found that the adsorption capacity decreased upon apatite maturation (table 6). On the other hand, a higher affinity for BSA was observed in matured samples, which was related to possible competition between water and BSA molecules for the interaction sites on the apatite crystals, and to the presence of non-apatitic phosphate groups which enhance electrostatic repulsion at the surface.

Table 6. Influence of maturation time on the adsorption of BSA onto nanocrystalline apatites in 1 mM KCl solution at pH ≅ 7.

Apatite samples	Maximum amount adsorbed (mmol/m^2)	Affinity constant (mL/mg)
Not matured	42.1 ± 5.3	0.5 ± 0.2
Matured 15 days	33.6 ± 1.2	39.0 ± 17.2
Matured 45 days	23.4 ± 0.7	229.9 ± 81.8

The protein adsorption ability of apatites can be used in the biomaterials field to improve the bone wound-healing processes by associating apatite (carrier) with growth factors which have been shown to be potent osteogenic inductors or with a specific protein RGD sequence (arginine-glycine-aspartic acid) involved in cell attachment to favor cell adhesion, proliferation and differentiation. Midy et al. reported a study on the adsorption and release properties of synthetic carbonated apatite (matured for 24h) analogous to bone mineral compared to the properties of well-crystallized hydroxyapatite toward basic fibroblast growth factor (bFGF) [Midy 1998]. The adsorption of bFGF was much greater on carbonated apatite (85% of the amount contained in solution) than on hydroxyapatite (10% of the amount contained in solution) probably due to the higher amount of non-apatitic environments of carbonate and hydrogenphosphate ions in carbonated apatite. Indeed it has been suggested that non-apatitic environments and proteins share the same adsorption sites at the surface of the apatite nanocrystals [Rey 1994]. However, the proportion of bFGF released remained low and quite the same for the two kinds of apatite (from 17 to 27 % of the adsorbed amount) probably due to strong interactions between bFGF and apatite surfaces through acidic residues of this growth factor (negatively charged) and also to the short duration of the experiment (60 min) related to the short growth factor half-life. This study showed that growth factor binding to apatites depends strongly on apatite characteristics (poorly crystallized non-stoichiometric apatites rich in non-apatitic surface species are more efficient than stoichiometric well-crystallized HA). In vivo, these properties could play a role in signaling the evolution of the mineral crystals (apatite) to cells and in bone remodeling.

It is interesting to note that interactions between biomimetic apatites and proteins can, in some conditions, lead to physico-chemical alterations of the initial solid phase. In this view, Luong et al. have recently investigated the effect of BSA on a biomimetically nucleated mineral, and showed that while surface adsorption of BSA did not lead to structural modifications of the mineral, BSA incorporation (occurring when added during the synthesis) changed the crystal morphology from plate-like to more rounded structures: similar observations have been made by Combes et al. in the case of OCP nucleation and crystal growth in the presence of BSA [Combes 2002, Luong 2006].

To gain a better understanding of the adsorption mechanisms on apatite nanocrystals, smaller molecules such as peptides or amino acids are often used for closer identification of possible interacting groups. It is thus generally accepted from experimental results that the alpha-carboxylate group of the amino acid is preferentially bound to the nanocrystal surface. The affinity was also found to be dependent on the apatite characteristics. For example in the case of glycine, the affinity decreased as the HPO_4^{2-} content decreased and as the carbonate content increased [Bennani-Ziatni 2003]. Also, the affinity was found to increase with the proportion of ethanol in the medium, i.e. with the decrease in dielectric constant. The effect of the presence of phosphate ions in the medium was investigated for alanine and phenylalanine adsorption [Bihi 2002]. It was then found that the phosphate ions hindered the adsorption process by competing with the amino acid carboxylate group for interacting with the calcium ions at the nanocrystal surface. Another interesting result arose from the comparison of adsorption parameters for serine and phosphoserine [Benaziz 2001]. In the latter case, a greater affinity was found, which was related to the presence of a phosphate group in this amino acid molecule, that has specific attachment sites. Evidence of the involvement of anionic groups of the amino acid in the adsorption process is in particular given by a shift of the related IR bands, stressing the close interaction with surface species on the crystals (e.g. shift for the carboxyl groups). Other functional groups of amino acids can however also be involved in the adsorption process. For example, the amine groups of glycine were also found to intervene [Bennani-Ziatni 1997].

Finally, the nature of the amino acid was found to play a role in the amount that can be bound on biomimetic apatites. Indeed, Bihi et al. studied the adsorption of alanine and phenylalanine on such compounds, and observed that the increase of the size of the amino acid side chain led to a decrease of maximum uptake at saturation [Bihi 2002].

Chapter 7

PROCESSING OF NANOCRYSTALLINE APATITE-BASED BIOMATERIALS

Nanocrystalline calcium phosphate apatites are interesting biomaterials due to their surface reactivity which has just started to be explored. The existence of a superficial hydrated layer including highly labile ions offers an opportunity to vary and adapt the surface properties, whether for materials science or for biological purposes. Thus, it is possible to take advantage of the reactivity of nanocrystalline calcium phosphate apatites in the processing of bioactive bone substitutes.

As illustrated in figure 15, one of the most interesting properties related to the existence of the hydrated surface layer is the ability of nanocrystals to join and strongly interact with each other and/or with different substrates and macromolecules. Interestingly, even though it is not yet fully understood, the junction of apatite crystals has been reported in living organisms for example during dental enamel formation and is referred to as "crystal fusion" [Daculsi 1984].

In the materials science field, the hydrated layer can be involved in cohesiveness and adhesion between two apatite nanocrystals or between apatite nanocrystals and different substrates in the low temperature processing of ceramics and coatings or composites respectively. The progressive drying increases inter-crystal or crystal-substrate contacts. Upon drying, the steady elimination of excess water molecules brings two crystals together enabling the constitutive ions to interact from a strong electrostatic interaction. At the end of the process, crystals that have been joined cannot be split apart by simple rehydration. Intrinsic and extrinsic parameters such as the extent of the hydrated layer at the surface of the nanocrystals (depending on the apatite maturation state) and the rate and temperature of drying, respectively, have to be considered to optimize the processing of nanocrystalline apatite-based biomaterials.

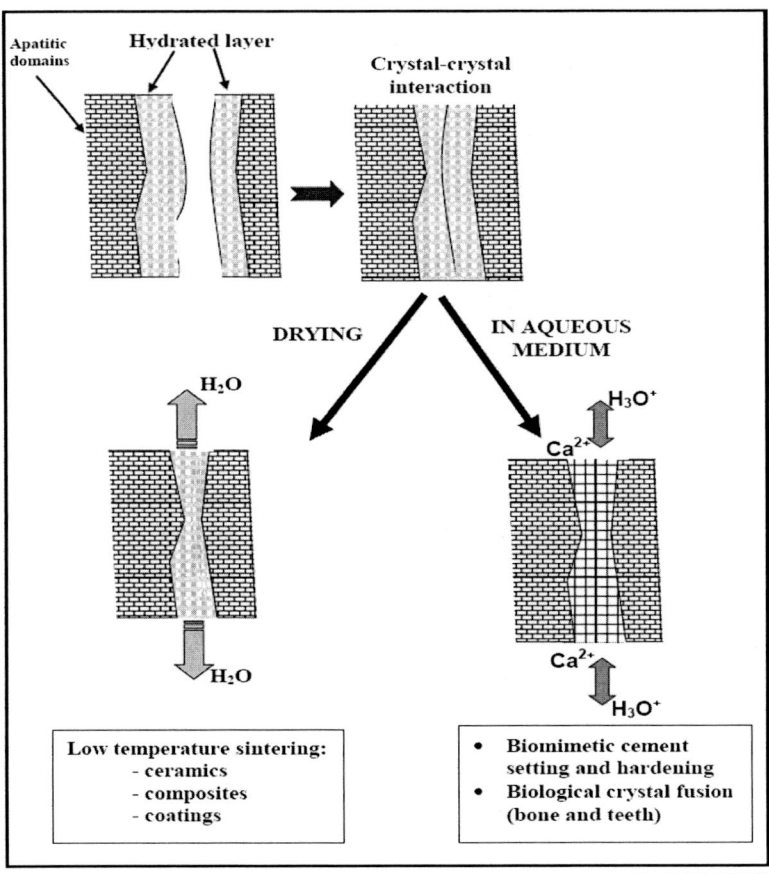

Figure 15. Role of the hydrated layer in the preparation of materials involving poorly crystalline apatite nanocrystals.

Depending on the type of repair, surgeons need bioceramics processed in different forms (dense or porous blocks, coatings, cements) in which apatites can be used alone, or as part of a composite, or formed as a result of a reaction in situ (bone cements). We present hereafter a short review of nanocrystalline apatite-based cements, coatings and composites biomaterials.

7.A. BIOMIMETIC CEMENTS

Biomimetic calcium phosphate (CaP) cements are based on the ability of CaP phases to form apatitic calcium phosphates in aqueous media. Different types of cements may be distinguished depending on the kind of reaction involved (table 7).

Table 7. Composition of biomimetic calcium phosphate cements.

Acid-base cements		
Composition	Final product	References
TTCP and DCPD	Hydroxyapatite (Bonesource®)	[Brown 1987]
$CaCO_3$+MCPM and α-TCP	Carbonated apatite (SRS®)	[Constantz 1995]
TTCP + H_3PO_4 and α-TCP	PCA (Cementek®)	[Hatim 1998]
$Ca(OH)_2$ or TTCP and OCP	PCA	[de Maeyer 2000]
Monocomponent cements		
Composition	Final product	References
amorphous CaP	Carbonated apatite (Biobon® and α-BSM®)	[Lee 2000]
α-TCP		[Ginebra 1997, Kon 1998, Dos Santos 1999]
Hydrolyzable phase-based cements		
Composition	Final product	References
β-TCP and MCPM	DCPD converting into PCA	[Mirtchi 1989]

DCPD: dicalcium phosphate dihydrate, OCP: octacalcium phosphate, MCPM: monocalcium phosphate monohydrate, TCP: tricalcium phosphate, TTCP: tetracalcium phosphate.

Acid-Base Cements

Most cements rely on an acid-base reaction between calcium phosphate phases. The acidic phase can be either a soluble CaP salt such as monocalcium phosphate or even phosphoric acid, or an insoluble CaP salt at physiological pH like dicalcium phosphate dihydrate (DCPD). The alkaline phase is generally tetracalcium phosphate (TTCP) but other phases like calcium carbonate or even calcium oxide have also been proposed.

The setting reaction in these cements is often complex and difficult to control as each of the constituents can hydrolyse separately in aqueous media into apatite. The pH of the paste can undergo strong variations during the setting time from acidic to alkaline. The particle size is a critical parameter which determines the setting characteristics of the cement. Generally the setting reaction involves several intermediates such as DCPD and/or OCP [Hatim 1998]. A multitude of formulations may be proposed for these cements and they offer a wide range of possibilities and characteristics. From a thermodynamic point of view, the acid-base reaction is always exothermic although the release of heat is generally much less than that of the well-known poly-methylmethacrylate (PMMA) biomedical cements [Baroud 2006].

Monocomponent Cements

Monocomponent CaP cements are based on the fast hydrolysis of one calcium phosphate salt into apatite. In order to achieve the short setting times needed for orthopaedic applications, the hydrolysis reactions have to be fast and the CaP phases should be rather unstable. Amorphous CaP and alpha-tricalcium phosphate (α-TCP) have been proposed [Lee 2000, Ginebra 1997]. The monocomponent cements induce less pH variation during setting and there is a direct conversion into an apatitic phase.

Generally the rate of the hydrolysis reaction is determined by the temperature. For amorphous calcium phosphate, for example, an increase of temperature causes an increase in the rate of conversion into apatite nanocrystals thus, at 20°C the paste does not harden and it is only at body temperature that fast hardening is achieved. These cements can easily be injected [Knaack 1998].

A third class of CaP cements has been developed that consists of unstable phases at physiological pH that transform into apatite after implantation (see table 7, hydrolyzable phase-based cements). The most frequently encountered examples are based on dicalcium phosphate dihydrate (DCPD) [Mirtchi 1989]. These are acid-base cements and the setting reaction corresponds to the formation of DCPD, which possesses a crystallographic structure analogous to that of gypsum or plaster of Paris. DCPD is an unstable CaP phase at neutral pH, and transforms into apatite at physiological pH.

Sometimes both the first (acid-base) and second type (conversion of an unstable CaP phase) of setting reactions are involved. Additives may often be used in all classes of CaP cements to control the setting reaction or limit pH variations.

The hardening reactions probably involve nanocrystal surface interactions between these particles in addition to the conversion of the precursors into apatite.

Similar phenomena have been described in biological tissues and are referred to as "crystal fusion".

Recently, original compositions of calcium carbonate-calcium phosphate mixed cements involving the reaction between hydrogenphosphate and carbonate groups have been developed [Combes 2006]. Depending on the initial carbonate/phosphate ratio, these cements can lead to a highly-carbonated nanocrystalline apatite (around 10% w/w of CO_3) analogous to bone mineral associated with a metastable calcium carbonate (vaterite) or not.

7.B. Low-Temperature Sintering of Nanocrystalline Apatites

As a preliminary, we can specify what is considered as "low temperature", since this appreciation appears to be completely dependent on the scientific context. Although ceramists, for example, generally consider temperatures below 1000 °C as being low, the notion of low temperature in this chapter will be reserved to heat treatments enabling the apatite samples to remain nanosized and hydrated, therefore typically in the range 20-300 °C. In these conditions, the term "consolidation" is then probably more suited than "sintering" to describe the physical phenomena undergone by the initial powder samples during the heat treatment.

Calcium phosphate ceramics exhibit excellent biocompatibility and show osteoconductive properties. As such, they are widely used as bone substitute materials and replace autologous grafts or allografts. However, these ceramics are generally obtained by sintering at high temperature and they therefore exhibit very limited surface reactivity and properties. As mentioned above, nanocrystalline apatites offer interesting possibilities: they are the main constituent of bone and nature makes good use of the particularities of their surface and bulk reactivity [Cazalbou 2004]. They are also considered to be at the origin of the biological activity of orthopaedic materials (bioglasses, polymers, ceramics or cements). However their processing while preserving their nanocrystalline nature is a major difficulty in view of the development of bioceramics made of apatite nanocrystals.

In this context, several explorative methods have been attempted such as low temperature hardening from gels, associations with macromolecules or cements [Knaack 1998, Sarda 1999]. However in these cases the mechanical properties observed were too weak to allow a use in load bearing bone sites.

An interesting feature of apatite nanocrystals, as explained in a previous section, is the presence of a surface hydrated layer, and Banu et al. recently investigated the ability of this hydrated layer to favor crystal-crystal interactions [Banu 2005]. They open interesting perspectives in materials science, since ceramic-like materials can be obtained upon drying aqueous gels of nanocrystals [Sarda 1999]. This process is however accompanied by strong shrinkage, which introduces strains in the materials leading to poor mechanical properties. The possibility to use the ion mobility of the hydrated layer to consolidate a nanocrystal assembly has however been established. Recently solid ceramic-like materials have been obtained at very low temperature by low temperature uni-axial pressing [Banu 2005]. The materials obtained at 200°C compared well, regarding mechanical properties, with those obtained by traditional sintering at much higher temperatures (1100-1250 °C), despite a lower densification ratio. At this low temperature slight crystal growth of the apatite domains occurred associated with hydroxylation and a partial decomposition into anhydrous dicalcium phosphate (DCPA).

Beside conventional sintering techniques and hot pressing, spark plasma sintering (SPS) is a technique recently used for the sintering of various kinds of materials at temperatures generally lower than usual, due to the non-conventional heating source (electrical current passing through a conducting matrix). Only few studies of apatite sintering using SPS have been performed, and most of them were done at relatively high temperature (900-1100 °C), not suited to the conservation of the nanosize of apatite crystals analogous to bone mineral. In these reports, SPS sintering was shown to be more efficient when HA powders with a small particle size were used, and the compacts exhibited excellent mechanical properties [Gu 2002]. The formation of transparent apatite ceramics with interesting surface properties has also been reported and generally SPS-sintered HA show greater surface reactivity in SBF (simulated body fluid) tests, often considered as an measurement of the biological activity [Nakahira 2002, Kawagoe 2003].

The process of consolidation of apatite nanocrystals at very low temperature using SPS and the specificity of apatite nanocrystals has not yet been studied in detail. Preliminary results confirm however the possibility to consolidate efficiently such nanocrystals at much lower temperatures than traditional sintering with limited degradation, which opens new fields in the making of ceramic-like biomaterials inspired by the bone mineral using the surface properties of hydrated nanocrystals, and more specifically the high ionic mobility within the surface hydrated layer present on the nanocrystals [Drouet 2006].

For example, consolidation by SPS of biomimetic nanocrystalline apatites matured for one day was investigated, and the spontaneous densification of the powder started around 150 °C and ranged up to 190 °C [Drouet 2006]. Disks obtained by SPS at 180 °C did not exhibit sufficient mechanical resistance, in contrast to those obtained around 200 °C. For the latter, the tensile stress measured by diametral compression test, or "Brazilian" test (adapted to the disk shape) ranged from 18 to 25 MPa for a heating time of 3 minutes from room temperature to 200 °C followed by a plateau at 200 °C for 2 minutes. These values are close to those obtained with stoichiometric hydroxyapatite using the same mechanical test procedure. The corresponding relative density was ca. 70%. Preliminary results also point out the correlated effect of temperature and mechanical pressing for SPS consolidation of nanocrystalline apatites, and a high apparent cohesion of the original agglomerates can be observed by scanning electron microscopy (SEM), which is in agreement with the relatively good mechanical properties.

The XRD pattern of the consolidated disks obtained by SPS treatment for 2 minutes at 200 °C corresponds to a poorly crystalline apatite, although an increase in the degree of crystallinity after SPS treatment is noted. No other crystalline phase is discerned. The average crystallite dimensions were estimated applying Scherrer's formula to lines (002) and (310), leading to a length close to 260 Å and a width of about 80 Å as compared to 220 and 55 Å, respectively, for the starting powder (see section 3.B). These trends point out an increase of the crystallite sizes due to SPS treatment, however the increase of the crystallite length is noticeably lower than that observed after hot pressing treatment, even at low temperature: 295 Å for a treatment at 200 °C for 15 min. The SPS process therefore only causes limited alteration of the initial nanocrystalline powder.

FTIR analysis of SPS consolidated nanocrystalline apatites indicates partial loss of water and the appearance of sharp absorption bands characteristic of apatitic OH⁻ vibrations. The formation of OH⁻ ions can be interpreted by the internal hydrolysis of some PO_4^{3-} ions as suggested by Heughebaert [Heughebaert 1982]:

$$PO_4^{3-} + H_2O \rightarrow HPO_4^{2-} + OH^- \qquad \text{(eq. 6)}$$

However, the water loss observed after SPS consolidation is less than that observed after hot pressing processes, even at low temperature (200 °C for 15 min). Also, secondary phases (e.g. pyrophosphates) sometimes observed after hot pressing are not observed by FTIR spectroscopy after SPS treatment [Banu

2005]. This phase purity could represent another advantage of SPS consolidation of nanocrystalline apatites.

7.C. APATITE COATINGS

The main subject of concern with the well-known metallic prostheses is the interface between the surrounding bone and the implant surface. During the last two decades various coating methods leading to hydroxyapatite (HA)-coated prostheses combining the good mechanical properties of metals with the excellent biocompatibility and bioactivity of calcium phosphate have been studied. HA-coating is known to improve bone formation when in contact with bone tissue (osteoconduction) and to facilitate the anchorage between the bone tissue and the prosthesis (biointegration). Currently, HA plasma-spraying is the most common and successful process to enhance the bioactivity of metallic implants (orthopaedic and dental implants) and to improve early bone-implant bonding. However, despite its clinical success, the plasma spray process is limited by intrinsic drawbacks: a) the high temperatures involved in this process restrict the process to the deposition of stable phases like stoichiometric hydroxyapatite presenting a low specific surface area (around 1 m^2/g) and thus a low reactivity without any possibility to associate biologically active (macro)molecules (growh factors for example) with the coating, b) the limit of this technique to cover complex surfaces and/or inside porous materials, c) the partial and superficial decomposition of HA particles into several possible phases including CaO, tricalcium phosphate, tetracalcium phosphate and oxyapatite at very high temperatures which could modify the chemical, mechanical and biological behavior of the biomaterial.

Other processes have been studied to improve the quality of the coatings (electrophoretic deposition, ion sputtering deposition and sol-gel methods for example) but in all cases the phase(s) composition of the coating is quite different from biological apatites and the coating reactivity is limited. For example, even though promising cell culture results were obtained (improved cell differentiation compared to pure titanium), one of the main problems of the apatite coating obtained by sol-gel route is the aging time of the precursor sol which can influence the phase composition, homogeneity and textural properties of the deposit [Kim 2004, Izquierdo-Barba 2004].

Recently, other deposition processes involving calcium phosphate supersaturated solution(s) at low temperature have been studied. Several authors explored the concept of biomimetic coatings which consists in soaking metal or polymer implants at physiological pH and temperature in Simulated Body Fluids solution (SBF) mimicking the inorganic ion composition of human blood plasma [Kokubo 1996, Barrère 1999, Leonor 2004]. The main advantages of these processes based on the nucleation and growth of calcium phosphate on metallic substrates from supersaturated solution arise from a) the low processing temperature, b) the formation of a non-stoichiometric apatite quite analogous to bone mineral, c) the possibility to form deposits even on complex implant geometries and to incorporate biologically-active molecules or ions in the coating. However, several drawbacks make this technique difficult to apply on an industrial scale: a) the long immersion time (problem of maturation of apatite and thus a decrease of its reactivity), b) the thinness of the apatite layer deposited and c) the metastability of SBF solutions, and more generally of calcium phosphate supersaturated solutions, which require replenishment and a constant pH to maintain the level of supersaturation for apatite precipitation. Habibovic et al. reported the deposition of an AB type carbonated hydroxyapatite involving two steps using two modified-SBF solutions: firstly, the heterogeneous nucleation of a thin and amorphous CaP layer on the metal surface and secondly the growth of a thick well-crystallized apatite coating [Habibovic 2002]. They have shown that drugs and/or growth factors could be easily incorporated into the biomimetic coating during processing in order to improve the biological performance of the biomaterial (prevention of local infection, controlled drug delivery, enhancement of bone healing).

To avoid the use of supersaturated solutions in industrial processing, another process involving the slow hydrolysis of apatite precursors such as amorphous calcium phosphate first deposited on the metallic surface has been set up [Cazalbou 2002].

On a hydrophilic surface such as the hydrated titanium oxide-hydroxide layer at the surface of a titanium implant in aqueous media, the interaction between apatite nanocrystals and the titanium surface can be achieved through the hydrated surfaces interacting with a bonding process as illustrated in figure 15. The labile ions from the superficial hydrated layer of freshly formed apatite allow a topological fitting to the metallic surface. The progressive transformation of the superficial layer into regular apatite structure improves the adhesion of the nanocrystals to the surface. Upon drying strong bonding may be formed by direct ionic interactions. Even though no systematic studies have been reported, several parameters involved in the quality of the bonding can be distinguished: the drying rate and/or the rate of evolution of the hydrated layer toward regular apatite structure.

Another application that takes advantage of the PCA reactivity is the activation of three-dimensional porous scaffolds often used for bone regeneration (bone filling and substitute, tissue engineering) by coating their accessible surfaces with biomimetic nanocrystalline apatites analogous to bone mineral. Indeed, although generally based on calcium phosphate phases (e.g. biphasic calcium phosphate ceramics (BCP): HA/β-TCP), the processing of such scaffolds usually leads to a strong decrease of their bioactivity due to high-temperature treatments and the coating of the ceramic open pores with PCA make it possible to take advantage of the physical-chemical properties of such nanocrystalline apatite phase to enhance the ceramic bioactivity. Figure 16 shows a poorly crystalline apatite coating (around 3 μm-thick) obtained by impregnation by PCA gel on the accessible surface of a porous BCP ceramic. Further activation can also be achieved by surface ion exchange between the biomimetic apatite coating and aqueous solutions enriched with mineral ions such as magnesium and carbonate (see in section 5.B).

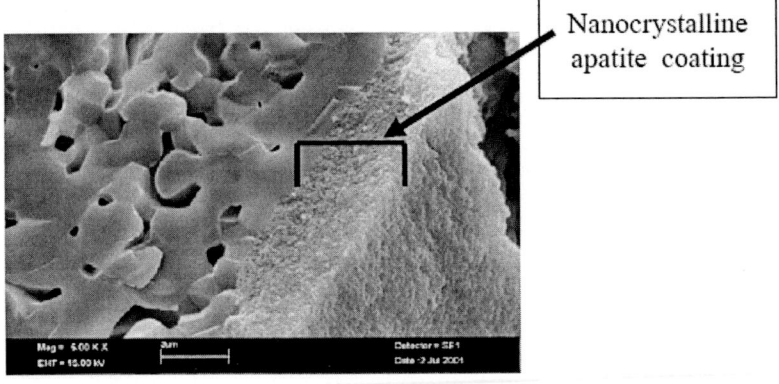

Figure 16. SEM micrograph of nanocrystalline apatite deposit on porous HA/TCP scaffold accessible surface.

7.D. COMPOSITES

The model of interactions between platelets of apatite nanocrystals and collagen (protein) occurring in bone tissue, although not yet completely understood, can be exploited to prepare hybrid mineral-organic materials. The association of calcium phosphates with organic molecules (mineral-organic composites) can impart bioactive and mechanical properties closer to those of bone compared to pure calcium phosphate ceramics. The possibility for molecules to interact with the ions from the apatite surface layer through ionic functions of proteins (mostly anionic groups) has already been mentioned in this chapter (see in section 5c). Polymer- or protein-apatite composites can combine a better structural integrity and flexibility along with good bioactivity. Recently, such composites have gained increasing interest in the field of tissue engineering which appears as a promising concept for bone reconstruction. This concept requires a biodegradable host matrix (composite for example) acting temporarily as a mechanical support and directing osteoblast growth and tissue neoformation once implanted. High porosity is necessary to promote implant binding and integration and especially neoformed bone invasion.

Generally, composite materials (ceramic-polymer) are obtained with classical mixing techniques of constituents but other techniques inspired by in vivo hard tissue calcification processes have emerged. For example, the use of organized organic matrices with numerous sites favourable for nucleation of calcium phosphates after phosphatation (of cellulose for example) or sodium silicate treatments [Miyaji 1999].

Many studies can be found in the literature pursuing the aim to produce biomimetic artifcial bone-like tissue involving HA and collagen as fiber, gel or gelatin (denatured collagen) [Itoh 2000, Tampieri 2003, Kikuchi 2004, Kim 2005]. Testing two methods of preparation of apatite/collagen composite materials (dispersion of HA in collagen gel or direct nucleation of HA into collagen fibers), Tampieri et al. have shown that the bio-inspired method based on the direct nucleation of apatite leads to composites analogous to calcified tissue and exhibiting strong interactions between HA and collagen [Tampieri 2003].

In the dog, in vivo evaluation in weight-bearing sites testing apatite/collagen composites prepared by co-precipitation method has shown interesting results especially when associated with recombinant human bone morphogenetic protein 2 (rhBMP-2) [Itoh 2002]. The controlled release of rhBMP-2 from the implant facilitates early formation of calli and hence new bone enabling early weight bearing.

Nishikawa et al. have examined the biodegradation of HA/collagen composites implanted in dogs by a tissue labeling method; they showed that this novel composite is useful for bone augmentation and that the calcium in the newly formed bone might have been released from the implant [Nishikawa 2005].

The association of apatite with chitosan (a natural polysaccharide obtained by deacetylation of chitin) is expected to give interesting composites for tissue engineering applications due to the biocompatibility, biodegradability and bioactivity properties of chitosan. Several authors have reported the preparation of porous chitosan-apatite composites using co-precipitation methods [Kong 2005, Rusu 2005]. The association of chitosan with apatite improved the biocompatibility of the material and greater cell proliferation on a composite scaffold has been reported [Kong 2005].

In our research group, protein-CaP associations have been prepared according to biomimetic processes at low temperature and in aqueous suspension. The proteins studied (casein, albumin) present an affinity for calcium phosphate surfaces. After evaporation of water (according to figure 15 drying scheme, "low temperature sintering" process) in the protein-CaP suspension, nanocrystalline apatite associated with protein gives a bulk composite (protein-nanocrystalline apatite) with low porosity but with poor mechanical properties compared to those of bone [Sarda 1999].

Macroporous composite scaffolds can be processed using different methods: solvent casting/particulate leaching, emulsion freeze drying or thermally induced phase separation.

Biodegradable composite including resorbable polymer such as polylactic acid (PLA) and/or polyglycolic acid (PGA) and a resorbable apatite can be prepared at ambient temperature [Linhart 2001, Chen 2001]. This kind of association has the advantage of a controlled pH in the surrounding medium during the composite degradation since apatite (whose degradation involves the release of alkaline components) can moderate the pH drop resulting from polyester (PLA or PGA) biodegradation (hydrolysis of ester bond). Other good reasons for using such synthetic or natural polymers are: a) the ease of processing , b) the possibility to control the polymer degradation rate depending on its composition (polymer or copolymer), molecular weight and crystallinity. Moreover, the degradation in vivo of PLA releases lactic acid, which is a natural metabolite and thus remains under the control of the organism. The biomaterial structure (dense or porous composite) also has an effect on the rate of biodegradation which is lower in the case of porous composites where the acid released from PLA degradation can be leached with biological fluids whereas this acid concentrated in the bulk of dense composite should autocatalyze composite biodegradation.

Recently, Mathieu et al. reported the use of a supercritical CO_2 foaming process to prepare porous PLA-HA and/or PLA-TCP composites exhibiting mechanical behavior analogous to bone (anisotropy in compressive and viscoelastic properties) [Mathieu 2006].

Finally, the formation of a polymer-apatite composite can also correspond to a first step in the preparation of nanocrystalline apatite porous ceramic. For example, Tadic et al. reported the preparation of nanocrystalline apatite-based porous bioceramics using both sodium chloride salt and polyvinyl alcohol fibers as water-soluble porogenic agents and cold isostatic pressing without the need to sinter [Tadic 2004].

Chapter 8

BIOLOGICAL PROPERTIES OF NANOCRYSTALLINE APATITES

Three main characteristics have to be considered: biodegradation, cell-materials interactions and materials-tissue interactions.

8.A. BIODEGRADATION

The main advantage of apatite nanocrystals-based biomaterials is their similarity with bone mineral. However these nanocrystals may have very different characteristics which account for the large differences in resorption rates. Several factors involved in biodegradation have been identified.

The composition of CaP nanocrystals appears as an important factor. Resorption is generally believed to be related to the solubility of CaP materials [Legeros 1993]. Considering nanocrystalline apatites, the solubility depends on the presence of vacancies and on the molar Ca/(P+C) ratio (where P represents phosphate ions and C the carbonate ions). Low Ca/(P+C) ratios indicate a high amount of cationic vacancies and a less cohesive solid with a higher solubility. Non-stoichiometric apatites, however, have the ability to mature like bone mineral and may evolve toward more stable, less soluble compounds upon aging. Thus a change in solubility may occur with time and the rate of bioresorption may decrease. These evolutions and alterations of the nanocrystal composition with implantation time have not been studied extensively, but will be crucial for understanding the degradation behavior of the nanocrystals.

The amount of labile ("non-apatitic") species may also considerably modify the dissolution characteristics and thus the resorption rate. It is likely that the labile species determine the "abnormal" dissolution characteristics of bone crystals [Baig 1999]. These factors depend on the composition of the materials and on their conditions of formation. Some materials, involving inhibitors of crystal maturation such as carbonate, magnesium or pyrophosphate, consist of small apatitic crystals rich in labile non-apatitic environments that will be rapidly resorbed. Others contain bigger, more mature crystals with a slower resorption rate. The presence of alkaline residues or the existence of alkaline reactions on contact with body fluids may strongly disturb biodegradation by cells as slight alkalinity was shown to greatly reduce osteoclast resorption [Kim 2001].

In addition to specific characteristics of the nanocrystals, their bonding scheme and the characteristics of the materials (porosity, size, presence of other constituents) also play a considerable role in the biodegradation process as for other types of biomaterials.

Although composed of apatite nanocrystals analogous to bone mineral, the degradation of biomimetic apatite-based materials may also produce small particles that can induce a local inflammatory response.

8.B. Biological Activity

The origin of biological activity has not really been investigated in all cases. Most studies concern implantations in animals, but the behavior of osteoblasts on CaP biomimetic materials is not well known. Several studies performed for example on cement particles have shown an adverse effect on osteoblast viability, proliferation and synthesis of extracellular matrix [Pioletti 2000]. These effects however, are not due to the cement itself but rather to the presence of loose particles and in fact may be observed with any kind of material releasing small particles. Generally, all implantation tests of biomimetic CaP materials have given excellent results. There is generally no formation of fibrous tissue at the contact of the implant, and the material is directly attached to bone. In most cases, the irregular resorption zone is invaded by new forming bone and osteons can often be distinguished as in regular bone remodeling. A few islands of materials are imbedded in new forming bone at the end of the process [Frankenburg 1998, Knaack 1998, Yuan 2000].

Some biomimetic cements and coatings have been reported to be osteoinductive. However the manifestations of this behavior seem rather erratic. Although assigned to microporosity and the preferential uptake of circulating growth factors, the origin of this behavior needs to be demonstrated and clarified [Habibovic 2006].

The reasons for biological activity and its correlation with the characteristics of the biomimetic material are difficult to extract from existing data. Nevertheless knowing which biomaterial-tissue interactions occur allows a probable scenario to be sketched. The bioactivity of orthopaedic materials has been shown to result from the precipitation of a neoformed layer of carbonated apatite crystals analogous to bone mineral crystals in vivo, which enables osteoblast cell adhesion and proliferation. The formation of this neo-formed apatite layer can be induced either by nucleation of crystals from supersaturated biological fluids or by induction of nucleation resulting from increases in the local concentrations of Calcium, phosphate, carbonate or hydroxide ions. The first process might not necessarily occur in the presence of non-stoichiometric apatite nanocrystals which show a higher solubility than stoichiometric apatite and would be in quasi-equilibrium with body fluids. However the nanocrystals with their very high specific surface area offer numerous crystal growth sites which can promote the formation of neoformed crystals. The second process may be induced by the presence of other components. In several cases unstable phases existing in the materials may play a crucial role by locally increasing the mineral ion concentrations favoring additional precipitation of very reactive apatite from body fluids. These neoformed crystals and/or the exposed biomimetic nanocrystals of the material itself exhibit a very high specific surface area and a high surface reactivity due to the presence of labile, non-apatitic environments of mineral ions like in freshly formed bone crystals. They can easily bind proteins including growth factors which can favor cell attachment, proliferation and activity. The behavior of biomimetic materials may thus be rather complex. It can also vary with the time of implantation, the porosity and the size of the specimen. Proteins and cells will not diffuse through long distances. As maturation of the nanocrystalline apatitic phases progresses, biological activity, like resorption ability, can vary with time and with the size of the bone defect. These phenomena however, may also be true for other types of ceramic implants and are not specific to the behavior of biomimetic materials. Several additives may be used to delay maturation and preserve the activity of inner surfaces.

8.C. MATERIALS-TISSUE INTERACTIONS AND BIOINTEGRATION

In addition to tissue reconstruction around and within implants, biomaterials can bind directly to living tissue without the interposition of a fibrous layer. The bone-bonding ability is determined by mechanical interlocking and chemical interactions. The first effect is related to the general geometry of the samples and has little to do with its composition. Chemical interactions are believed to mainly involve the neo-formed apatite layer, however in the case of biomimetic nanocrystals the surface characteristics are already similar to that of bone tissue. Although bone bonding processes are not yet clearly known, it seems that the junction zone between the implant and bone tissue shows similar characteristics to the cement line separating old osteons from new osteons after bone remodelling: a disordered area with specific proteins such as ostepontin and bone sialoproteins [Kawagushi 1993, Davies 2003]. Thus, the bonding of living bone to biomimetic nanocrystalline apatite materials could be very similar to that occuring in bone remodelling processes. This binding process could explain the strength of the bone-apatite biomaterials interface. The interactions at the molecular level are not yet precisely known but it seems probable that the very reactive surface of bone nanocrystals could be involved in interactions both with the mineral fraction and the organic fraction.

CONCLUSION

Biomimetic nanocrystalline apatites exhibit enhanced and tunable reactivity as well as original surface properties related to their composition and mode of formation. Synthetic nanocrystalline apatites analogous to bone mineral can be easily prepared in aqueous media and one of the most interesting characteristics of their nanocrystals is the existence of a hydrated surface layer containing labile ionic species. The latter can be easily and rapidly exchanged with ions and/or macromolecules from the surrounding fluids. Even though the surface composition and structure of apatite nanocrystals have not yet been completely unveiled, the fine characterization of these very reactive nanocrystals can be investigated in different ways including chemical analysis, X-ray diffraction and spectroscopic techniques such as FTIR and solid state NMR. All these investigations are also essential to throw light on bone mineral reactivity and properties in general.

The ion mobility in the hydrated layer allows direct crystal-crystal or crystal-substrate bonding offering material scientists and medical engineers extensive possibilities for the design of bone substitute materials with improved bioactivity using unconventional processing. Indeed apatitic biomaterials can be processed in different forms (dense or porous ceramics, cements, composites and coatings) at low temperatures, which preserves their surface reactivity and biological properties. They can also be associated with bioactive (macro)molecules and/or ions to improve the biological performance of bone substitute materials.

REFERENCES

Arsenault A.L., Grynpas M.D. (1988). Crystals in calcified epiphyseal cartilage and cortical bone of the rat. *Calcified Tissue International*, 43, 219-225.

Aue W.P., Roufosse A.H., Glimcher M.J., Griffin R.G. (1984). Solid state phophorus-31 Nuclear Magnetic Resonance of calcium phosphate: Potential models of bone mineral. *Biochemistry,* 23, 6110-6114.

Baig A.A., Fox J.L., Hsu J., Wang Z., Otsuka M., Higuchi W.I., Legeros R.Z. (1996). The effect of carbonate content and crystallinity on the metastable equilibrium solubility behavior of carbonated apatites. *J Colloid Interface Sci.*, 179, 608–617.

Baig AA, Fox JL, Wang Z, Higuchi WI, Miller SC, Barry AM, Otsuka M. (1999). Metastable equilibrium solubility behavior of bone mineral. *Calcif. Tissue Int.,* 64, 329-339.

Baig A.A., Fox J.L., Young R.A., Wang Z., Hsu J., Higuchi W.I., Chhettry A., Zhuang H., Otsuka M. (1999). Relationships among carbonated apatite solubility, crystallite size, and microstrain parameters. *Calcified Tissue International*, 64, 437–449.

Banu M. Ph.D Thesis, INPT, France, 2005.

Baroud G., Swanson T., Steffen T. (2006). Setting properties of four acrylic and two calcium phosphate cements used in vertebroplasty. *J. Long Term Eff. Med. Implants*, 16, 51-59.

Barrère F., Layrolle P., van Blitterswijk C.A., de Groot K. (1999). Biomimetic calcium phosphate coatings on Ti6AI4V: a crystal growth study of octacalcium phosphate and inhibition by Mg^{2+} and HCO_3^-. *Bone*, 25, 107S-111S.

Belton P.S., Harris R.K., Wilkes P.J. (1988). Solid state phophorus-31 NMR studies of synthetic inorganic phosphates. *J. Phys. Chem. Solids*, 49, 21-27.

Benaziz L., Barroug A., Legrouri A., Rey C., Lebugle A. (2001). Adsorption of O-phospho-L-serine and L-serine onto poorly crystalline apatite. *J. Colloid and Interface Science*, 238, 48-53.

Bennani-Ziatni M., Lebugle A., Taitai A., Ferhat M. (1997). Adsorption of glycine from a water-ethanol medium onto precipitated calcium phosphates. *Annales de Chimie*, 22, 537-547.

Bennani-Ziatni M., Lebugle A., Taitai A. (2003). Influence of dielectric constant on the adsorption of glycine by bioapatites. *Phosphorus, Sulfur and Silicon and the Related Elements*, 178, 221-233.

Beshah K., Rey C., Glimcher M.J., Schimizu M., Griffin R.G. (1990). Solid state carbon-13 and proton NMR studies of carbonate-containing calcium phosphates and enamel. *J. Solid State Chem.*, 84, 71-81.

Bihi N., Bennani-Ziatni M., Taitai A., Lebugle A. (2002). Adsorption of aminoacids onto bone-like carbonated calcium phosphates, *Annales de Chimie*, 27, 61-70.

Bonar L.C., Roufosse A.H., Sabine W.K., Grynpas M.D., Glimcher M.J. (1983). X-ray diffraction studies of the crystallinity of bone mineral in newly synthesized and density fractionated bone. *Calcified Tissue International*, 35, 202-209.

Bonfield W. (1988). Composites for bone replacement. *J Biomed Eng*, 10, 522-526.

Brown W.E., Chow L.C. (1987). A new calcium phosphate water-setting cement. In P Brown (Ed.), Cement Research Progress 1986, pp. 351-379. Westerville OH: *American Ceramic Society*.

Brown W.E., Eidelman N., Tomazic B. (1987). Octacalcium phosphate as a precursor in biomineral formation. *Adv. Dental Res.*, 1, 306-313.

Burnell J.M., Teubner E.J., Miller A.G. (1980). Normal maturational changes in bone matrix, mineral and crystal size in the rat. *Calcified Tissue International*, 31, 13-19.

Campbell A.C., LoRe M., Nancollas G.H. (1991). The influence of carbonate and magnesium ions on the growth of hydroxyapatite and carbonated apatite and human powdered enamel. *Colloids and Surfaces A*, 54, 25-31.

Cazalbou S., Hina A., Rey C. (1999). Interactions between trace elements and bone mineral matrix. In M. Abdulla, M. Bost, S. Gamon, P. Arnaud, G. Chazot (Eds.), *New Aspects of Trace Element Research*, pp. 58-62. London, Smith-Gordon.

Cazalbou S., Ph.D. Thesis, INPT, France, 2000.

Cazalbou S., Combes C., Rey C. (2002). Procédé permettant de recouvrir à basse température des surfaces par des phosphates apatitiques nanocristallins à partir d'une suspension aqueuse de phosphate amorphe. Patent n° FR0209514.

Cazalbou S., Combes C., Eichert D., Rey C., Glimcher M.J. (2004). Poorly crystalline apatites: evolution and maturation in vitro and in vivo. *J Bone Miner Metab*, 22, 310-317.

Cazalbou S., Combes C., Eichert D., Rey C. (2004). Adaptative physico-chemistry of bio-related calcium phosphates. *Journal of Materials Chemistry*, 14, 2148-2153.

Chen G., Ushida T., Tateishi T. (2001). Poly(DL-lactic-co-glycolic acid) sponge hybridized with collagen microsponges and deposited apatite particulates. *J Biomed Mater Res*, 57, 8-14.

Cho G., Yesionowski J.P. (1996). ^{1}H and ^{19}F Multiple-Quantum NMR dynamics in quasi-one-dimensional spin clusters in apatites. *J. Phys. Chem.*, 100, 15716-15725.

Cho G., Wu Y., Ackerman J.L. (2003). Detection of hydroxyl ions in bone mineral by solid-state NMR spectroscopy. *Science*, 300, 1123-1127.

Combes C., Rey C., Mounic S. (2001). Identification and evaluation of HPO_4 ions in biomimetic poorly crystalline apatite and bone mineral. *Key Engineering Materials*, vol 192-195, 143-146.

Combes C., Rey C. (2002). Adsorption of proteins and calcium phosphate materials bioactivity. *Biomaterials*, 23, 2817-2823.

Combes C., Bareille R., Rey C. (2006). Calcium carbonate-calcium phosphate mixed cement compositions for bone reconstruction. *J Biomed Mater Res A* (in press).

Constantz B.R., Ison I.C., Fulmer M.T., Poser R.D., Smith S.T., VanWagoner M., Ross J., Goldstein S.A., Jupiter J.B., Rosenthal D.I. (1995). Skeletal repair by in situ formation of the mineral phase of bone. *Science*, 267, 1796-1798.

Dabbarh F., Lebugle A., Taitai A., Bennani M. (2000). Influence du séchage sur la composition de phosphates de calcium carbonatés, analogues osseux. *Ann Chim. Sci. Mat.*, 25, 339-348.

Daculsi G., Menanteau J., Kerebel L.M., Mitre D. (1984). Length and shape of enamel crystals. *Calcif Tissue Int*, 36, 550-555.

Daculsi G., Laboux O., Malard O., Weiss P. (2003). Current state of the art of biphasic calcium phosphate bioceramics. *J. Mater. Sci. Mater. Med.*, 14, 195-200.

Danilchenko S.N., Kukharenko O.G., Moseke C., Protsenko I.Y., Sukhodub L.F., Sulkio-Cleff B. (2002). Determination of the bone mineral crystallite size and lattice strain from diffraction line broadening. *Cryst. Res. Technol.*, 37, 1234-1240.

Davies J.E. (2003). Understanding peri-implant endosseous healing. *J. Dent. Educ.*, 67, 932-949.

de Groot K., Geesink R., Klein C.P., Serekian P. (1987). Plasma sprayed coatings of hydroxyapatite. *J. Biomed. Mater. Res.*, 21, 1375-1381.

de Jong W.F. (1926). The mineral substance in bones. *Rec. Trav. Chim., Pays-Bas*, 45, 445-448.

de Maeyer E.A., Verbeeck R.M., Vercruysse C.W. (2000). Conversion of octacalcium phosphate in calcium phosphate cements. *J Biomed Mater Res*, 52, 95-106.

dos Santos L.A., de Oliveira L.C., Rigo E.C., Carrodeguas R.G., Boschi A.O., De Arruda A.C. (1999). Influence of polymeric additives on the mechanical properties of α-tricalcium phosphate cement. *Bone*, 25, 99S-102S.

Drouet C., Largeot C., Raimbeaux G., Estournès C., Dechambre G., Combes C., Rey C.
(2006). Bioceramics: spark plasma sintering (SPS) of calcium phosphates. *Adv Sci Tech*, 49, 45-50.

Eichert D. Ph.D. Thesis, INPT, France, 2001.

Eichert D., Sfihi H., Combes C., Rey C. (2004). Specific characteristics of wet nanocrystalline apatites. Consequences on biomaterials and bone tissue. *Key Engineering Materials*, vol. 254-256, 927-930.

Eichert D., Salomé M., Banu M., Susini J., Rey C. (2005). Preliminary characterization of calcium chemical environment in apatite and non-apatitic calcium phosphates of biological interest by X-ray absorption spectroscopy. *Spectrochimica Acta* B, 60, 850-858.

Elliott J.C., Mackie P.E., Young R.A. (1973). Monoclinic hydroxyapatite. *Science*, 180, 1055-1057.

Elliott J.C., Holcomb D.W., Young R.W. (1985). Infrared determination of the degree of substitution of hydroxyl by carbonate ions in human dental enamel. *Calcif. Tissue Int.*, 37, 372-375.

Elliott J.C. (1994). Structure and Chemistry of the Apatites and Other Calcium Orthophosphates. Amsterdam, Elsevier Science BV.

Fernandez Gavarron F. (1978). The dynamic equilibrium of calcium. In E. Pina, A. Pena, V. Chagoya de Sanchez (Eds.), *Temas Bioquim.* Actual., pp. 41-67. Mexico Cit, Univ. Nac. Auton. Mexico.

Fisher L.W., Eanes E.D., Denholm L.J., Heywood B.R., Termine J.D. (1987). Two bovine models of osteogenesis imperfecta exhibit decreased apatite crystal size. *Calcified Tissue International*, 40, 282-285.

Frankenburg E.P., Goldstein S.A., Bauer T.W., Harris S.A., Poser R.D. (1998). Biomechanical and histological evaluation of a calcium phosphate cement. *J Bone Joint Surg Am*, 80, 1112-1124.

Gee A., Dietz V.R. (1953). Determination of phosphate by differential spectrophotometry. *Ann. Chem.*, 25, 1320-1324.

Gee A., Dietz V.R. (1955). Pyrophosphate formation upon ignition of precipitated basic calcium phosphate. *J. Am. Chem. Soc.*, 77, 2961-2965.

Ginebra M.P., Fernandez E., De Maeyer E.A., Verbeeck R.M., Boltong M.G., Ginebra J., Driessens F.C., Planell J.A. (1997). Setting reaction and hardening of an apatitic calcium phosphate cement. *J Dent Res*, 76, 905-912.

Glimcher M.J. (1990). The possible role of collagen fibrils and collagen-phosphoprotein complexes in vitro and in vivo. *Biomaterials*, 11, 7-10.

Grynpas M.D., Bonar L.C., Glimcher M.J. (1984). X-ray diffraction radial distribution function studies of bone mineral and synthetic calcium phosphate. *J. Mater. Sci.*, 19, 723-736.

Gu Y.W., loh N.H., Khor K.A., Tor S.B., Cheang P. (2002). Spark plasma sintering of hydroxyapatite powders. *Biomaterials*, 23, 37-43.

Habibovic P., Barrère F., van Blitterswicjk C.A., de Groot K., Layrolle P. (2002). Biomimetic hydroxyapatite coating on metal implants. *J Am Ceram Soc*, 85, 517-522.

Habibovic P., Sees T.M., van den Doel M.A., van Blitterswijk C.A., de Groot K. (2006). Osteoinduction by biomaterials. Physicochemical and structural influences. *J. Biomed. Mater. Res.* A, 77, 747-762.

Haris Parvez I., Bigi A., Boanini E., Gazzano M., Rubini K., Torricelli P. (2004). Nanocrystalline hydroxyapatite-polyaspartate composites. *Bio-medical Materials and Engineering*, 14, 573-579.

Hatim Z., Frèche M., Kheribech A., Lacout J.L. (1998). The setting mechanism of a phosphocalcium biological cement. *Ann. Chim. Sci. Mat.*, 23, 65-68.

Heughebaert J.C., Montel G. (1982). Conversion of amorphous tricalcium phosphate into apatitic tricalcium phosphate. *Calcif Tiss Int*, 34, S103-S108.

Hina A. Ph.D. Thesis, INPT, France, 1996.

Huffman E.W.D. (1977). Performance of a new automatic carbon dioxide coulometer. *Microchemical J*, 22, 567-573.

Ishikawa K., Ducheyne P., Radin S. (1993). Determination of the calcium/phosphorus ratio in calcium-deficient hydroxyapatite using X-ray diffraction analysis. *J. Mat. Sci: Mater. In Med.*, 4, 165-168.

Isobe T., Nakamura S., Nemoto R., Senna M., Sfihi. H. (2002). Solid state double nuclear magnetic resonance of calcium phosphate nanoparticules synthesized by wet-mechanochemical reaction. *J. Phys. Chem* B, 106, 5169-5176.

Itoh S., Kikuchi M., Takakuda K., Koyama Y., Matsumoto H.N., Ichinose S., Tanaka J., Kawauchi T., Shinomiya K. (2000). The biocompatibility and osteoconductive activity of a novel hydroxyapatite /collagen composite biomaterial, and its function as a carrier of rhBMP-2. *J. Biomed Mater Res*, 54, 445-453.

Itoh S., Kikuchi M., Takakuda K., Nagaoka K., Koyama Y., Tanaka J., Shinomiya K. (2002). Implantation study of a novel hydroxyapatite /collagen (Hap/col) composite into weight-bearing sites of dogs. *J. Biomed Mater Res*, 63, 507-515.

Iyengar G.V., Tandon L. Minor and trace elements in human bones and teeth. *International Atomic Energy Agency, NAHRES-39 report*, Vienna,1999.

Izquierdo-Barba I., Hijon N., Cabanas M.V., Vallet-Regi M. (2004). Apatite layers by a sol-gel route. *Key Engineering Materials,* vol. 254-256, 363-366.

Jäger C., Welzel T., Meyer-Zaika W., Epple M. (2006). A solid state NMR investigation of the structure of nanocrystalline hydroxyapatite. *Magn. Reson. Chem.*, 44, 573-580.

Jarcho M., Salsbury R.L., Thomas M.B., Doremus R.H. (1979). Synthesis and fabrication of β-tricalcium phosphate ceramics for potential prosthetic applications. *J Mater Sci,* 14, 142-150.

Jarlbring M., Sandström D.E., Antzutkin O.N., Forsling W. (2006). Characterization of Active Phosphorus Surface Sites at Synthetic Carbonate-Free Fluorapatite Using Single-Pulse ^{1}H, ^{31}P, and ^{31}P CP MAS NMR. Langmuir, 22, 4787 - 4792.

Johnson A.R., Armstrong W.D., Singer L. (1970). Exchangeability of calcium and strontium of bone in vitro. *Calcified Tissue Research,* 6, 103-112.

Kawagoe D., Ioku K., Fujimori H., Goto S. (2003). In vitro estimation with simulated body fluid of OH-designed transparent apatite ceramics prepared by spark plasma sintering. *Trans. Mater. Res. Soc.* Japan., 28, 841-844.

Kawagushi H., McKee M.D., Okamoto H., Nanci A. (1993). Tissue response to hydroxyapatite implanted in an alveolar bone defect: Immunocytochemical and lecitin-gold characterization of the hydroxyapatite-bone interface. *Cells and Materials,* 3, 337-350.

Kikuchi M., Ikoma T., Itoh S., Matsumoto H.N., Koyama Y., Takakuda K., Shinomiya K., Tanaka J. (2004). Biomimetic synthesis of bone-like nanocomposites using the self-organization mechanism of hydroxyapatite and collagen. *Composites Science and Technology,* 64, 819-825.

Kim H.M., Kim Y.S., Woo K.M., Park S.J., Rey C., Kim Y., Kim J.K., Ko J.S. (2001). Dissolution of poorly crystalline apatite crystals by osteoclasts determined on artificial thin-film apatite. *J. Biomed. Mater. Res.,* 56, 250-256.

Kim H.W., Kim H.E., Knowles J.C. (2004). Sol-gel films on titanium implant for hard tissue regeneration. *Key Engineering Materials,* vol. 254-256, 423-426.

Kim H.W., Knowles J.C., Kim H.E. (2005). Porous scaffolds of gelatin-hydroxyapatite nanocomposites obtained by biomimetic approach: characterization and antibiotic drug release. *J Biomed Mater Res,* 74B, 686-698.

Knaack D., Goad M.E.P., Ailova M., Rey C., Tofighi A., Chakravarthy P., Lee D. (1998). Resorbable calcium phosphate bone substitute. *J. Biomed. Mat. Res.,* 43, 399-409.

Kokubo T. (1996). Formation of biologically active bone-like apatite on metals and polymers by a biomimetic process. *Thermochimica Acta,* 280/281, 479-490.

Kon M., Miyamoto Y., Asaoka K., Ishikawa K., Lee H.H. (1998). Development of calcium phosphate cement for rapid crystallization to apatite. *Dent Mater J,* 17, 223-232.

Kong L., Gao Y., Cao W., Gong Y., Zhao N., Zhang X. (2005). Preparation and characterization of nano-hydroxyapatite/chitosan composite scaffolds. *J Biomed Mater Res,* 75, 275-282.

Labarthe J.C., Bonel G., Montel G. (1973). Structure and properties of B-type phosphocalcium carbonate apatites. *Annales de Chimie* (Fr), 8, 289-301.

Lebugle A., Zahidi E., Bonel G. (1986). Effect of structure and composition on the thermal decomposition of calcium phosphates (Ca/P = 1.33). *Reactivity of Solids,* 2, 151-161.

Lee D., Rey C., Aiolova M., Tofighi A. (2000). Methods and products related to the physical conversion of reactive amorphous calcium phosphate, US patent N° 6117456.

Legeros R.Z., Trautz O.R., Legeros J.P., Klein E. (1968). Carbonate substitution in the apatitic structure. Bull. *Soc. Chim. Fr. n° special,* 1712-1718.

Legeros RZ. (1993). Biodegradation and bioresorption of calcium phosphate ceramics. *Clin. Mater.,* 14, 65-88.

Legeros R.Z. (1994). Biological and synthetic apatites. In P.W. Brown, B.Constantz (Eds.), *Hydroxyapatite and Related Materials,* pp. 3-28. Boca Raton, CRC Press.

Legeros R.Z. (2002). Properties of osteoinductive biomaterials: calcium phosphates. *Clin. Orthop. Relat. Res.,* 395, 81-98.

Legros R., Balmain N., Bonel G. (1987). Age-related changes in mineral of rat and bovine cortical bone. *Calcif. Tissue Int.,* 41, 137-144.

Leonor I.B., Azevedo H.S., Pashkuleva I., Oliveira A.L., Alves C.M., Reis R.L. (2004). Learning from nature how to design biomimetic calcium-phosphate coatings. In R.L. Reis and S. Weiner (Eds.), Learning from nature how to design new implantable biomaterials: from biomineralization fundamentals to biomimetic materials and processing routes, Nato Science Series; series II: *Mathematics, Physics and Chemistry* - vol. 171, pp. 123-150. Dordrecht, Kluwer Academic Publishers.

Leung Y., Walters M.A., LeGeros R.Z. (1990). Second derivative infrared spectra of hydroxyapatite. *Spectrochimica Acta A.*, 46, 1453-1459.

Linhart W., Peters F., Lehmann W., Schwarz K., Schilling A.F., Amling M., Rueger J.M., Epple M. (2001). Biologically and chemically optimized composites of carbonated apatite and polyglycolide as bone substitution materials. *J Biomed Mater Res*, 54, 162-171.

Liu D.M., Troczynski T., Tseng W.J. (2002). Aging effect on the phase evolution of water-based sol-gel hydroxyapatite. *Biomaterials*, 23, 1227-1236.

Luong L.N., Hong S.I., Patel R.J., Outslay M.E., Kohn D.H. (2006). Spatial control of protein within biomimetically nucleated mineral. *Biomaterials*, 27, 1175-1186.

MacConnel D. (1973). *Apatite, its Crystal Chemistry, Mineralogy, Utilization, and Geologic and Biologic Occurrences.* New York, Spinger-Verlag.

Marie P.J., Ammann P., Boivin G., Rey C. (2001). Mechanisms of action and therapeutic potential of strontium in bone. *Calcified Tissue International*, 69, 121-129.

Mathew M., Brown W.E., Schoeder L.W., Dickens B. (1988). Crystal structure of octacalcium bis(hydrogenphosphate) tetrakis(phosphate) pentahydrate, Ca_8 $(HPO_4)_2$ $(PO_4)_4 \cdot 5H_2O$. *J. Cryst Spectrosc. Res.*, 18, 235-250.

Mathieu L.M., Mueller T.L., Bourban P.E., Pioletti D.P., Muller R., Manson J.A.E. (2006). Architecture and properties of anisotropic polymer composite scaffolds for bone tissue engineering. *Biomaterials*, 27, 905-916.

Midy V., Rey C., Brès E., Dard M. (1998). Basic fibroblast growth factor adsorption and release properties of calcium phosphate. *J Biomed Mater Res*, 41, 405-411.

Miquel J.L., Facchini L., Legrand A P., Rey C., Lemaitre J. (1990). Solid state NMR to study calcium phosphate ceramics. *Colloids and Surfaces*, 45, 427-433.

Mirtchi A.A., Lemaitre J., Terao N. (1989). Calcium phosphate cements: study of the beta-tricalcium phosphate-monocalcium phosphate system. *Biomaterials*, 10, 475-480.

Miyaji F., Kim H.M., Handa S., Kokubo T., Nakamura T. (1999). Bonelike apatite coating on organic polymers: novel nucleation process using sodium silicate solution. *Biomaterials*, 20, 913-919.

Nakahira A., Tamai M., Aritani H., Nakamura S., Yamashita K. (2002). Biocompatibility of dense hydroxyapatite prepared using a SPS Process. *J. Biomed. Mater. Res.*, 62, 550-557.

Neuman W.F., Toribara T.Y., Mulryan B.J. (1956). The surface chemistry of bone. IX. Carbonate: phosphate exchange. *J. Am. Chem. Soc.*, 78, 4263-4266.

Neuman W.F., Terepka A.R., Canas F., Triffitt J.T. (1968). Cycling concept of exchange in bone. *Calcified Tissue Research*, 2, 262-270.

Neuman W.F., Neuman M.W. (1985). Blood: bone calcium homeostasis. Shika Kiso Igakkai Zasshi, 24, 272-281.

Nishikawa T., Masuno K., Tominaga K., Koyama Y., Yamada T., Takakuda K., Kikuchi M., Tanaka J., Tanaka A. (2005). Bone repair analysis in a novel biodegradable hydroxyapatite/collagen composite implanted in bone. *Implant Dent*, 14, 252-260.

Ouizat S., Barroug A., Legrouri A., Rey C. (1999). Adsorption of bovine serum albumin on poorly crystalline apatite: influence of maturation. *Mat. Res. Bull.*, 34, 2279-2289.

Pak C.Y.C., Bartter F.C. (1967). Ionic interaction with bone mineral. I. Evidence for an isoionic calcium exchange with hydroxyapatite. *Biochimica et Biophysica Acta*, General Subjects, 141, 401-409.

Pan Y. (1995). ^{31}P-^{19}F rotational-echo, double resonance nuclear magnetic resonance experiment on fuoridated hydroxyapatite. *Solid State Nuclear Magn. Reson.*, 5, 263-268.

Panda R.N., Hsieh M.F., Chung R.J., Chin T.S. (2003). FTIR, XRD, SEM and solid state NMR investigations of carbonate-containing hydroxyapatite nano-particles synthesized by hydroxide-gel technique. *J. Phys. Chem. Solids*, 64, 193-199.

Paschalis, E.P., DiCarlo E., Betts F., Sherman P., Mendelsohn R., Boskey A.L. (1996). FTIR microspectroscopic analysis of human osteonal bone. *Calcif. Tissue Int.*, 59, 480-487.

Pasteris J.D., Wopenka B., Freeman J.J., Rogers K., Valsami-Jones E., van der Houten J.A.M., Silva M.J. (2004). Lack of OH in nanocrystalline apatite as a function of degree of atomic order: implications for bone and biomaterials. *Biomaterials*, 25, 229-238.

Penel G., Leroy G., Rey C., Brès E. (1998). MicroRaman spectral study of the PO_4 and CO_3 vibrational modes in synthetic and biological apatites. *Calcif. Tissue Int.*, 63, 475-481.

Pioletti D.P., Takei H., Lin T., Van Landuyt P., Ma Q.J., Kwon S.Y., Sung K.L. (2000). The effects of calcium phosphate cement particles on osteoblast functions. *Biomaterials*, 21,1103-14.

Pors Nielsen S. (2004). The biological role of strontium. *Bone,* 35, 583-588.

Posner A.S., Perloff A, Dioro A.F. (1958). Refinement of hydroxyapatite structure. *Acta Cryst.,* 11, 308-309.

Rey C., Collins B., Goehl T., Dickson R.I., Glimcher M.J. (1989). The carbonate environment in bone mineral. A resolution enhanced Fourier transform infrared spectroscopy study. *Calcif Tissue Int,* 45, 157-164.

Rey C., Shimizu M., Collins B., Glimcher M.J. (1990). Resolution enhanced Fourier transform infrared spectroscopic study of the environment of phosphate ion in the early deposits of a solid phase of calcium phosphate in bone and enamel and their evolution with age : I-investigation in the v4 PO4 domain. *Calcif. Tissue Int.,* 46, 384-394.

Rey C., Strawich E., Glimcher M.J. (1994). Non-apatitic environments in Ca-P biominerals; implications in reactivity of the mineral phase and its interactions with organic matrix constituents. In D. Allemand, J.P. Cuif (Eds.), *Biomineralizations,* pp. 55-64. Bulletin Océanographique n° special 14 1, Monaco.

Rey C., Kim H.M., Glimcher M.J. (1994) Maturation of poorly crystalline synthetic and biological apatites. In P.W. Brown, B. Constantz (Eds.), *Hydroxyapatite and Related Materials*, pp. 181-187. Boca Raton, CRC Press.

Rey C., Hina A., Tofighi A., Glimcher M.J. (1995). Maturation of poorly crystalline apatites: chemical and structural aspects in vivo and in vitro. *Cells and Mater.,* 5, 345-356.

Rey C., Miquel J.L., Facchini L., Legrand A.P., Glimcher M.J. (1995). On the question of hydroxyl groups in bone mineral. *Bone,* 16, 583-586.

Rey C., Combes C., Drouet C., Sfihi H., Barroug A. (2006). Physico-chemical properties of nanocrystalline apatites: Implications for biominerals and biomaterials. *Materials Science and Engineering C* (in press).

Roberts J.E., Bonar L.C., Griffin R.G., Glimcher M.J. (1992). Characterization of very young mineral phases of bone by solid state [31]Phosphorus magic angle sample spinning nuclear magnetic resonance and X-ray diffraction. *Calcified Tissue Int.,* 50, 42-48.

Rodrigues A., Lebugle A. (1998). Influence of ethanol in the precipitation medium on the composition, structure and reactivity of tricalcium phosphate. *Colloid and Surfaces* A, 145, 191-204.

Rogers K., Etok S., Broadhurst A., Scott R. (2005). Enhanced analysis of biomaterials by synchrotron diffraction. *Nuclear Instruments and Methods in Physics Research,* Section A, 548, 123-128.

Rothewel W.P., Waugh J.S, Yesinowsk J.P. (1980). High-resolution variable temperature ^{31}P NMR of solid calcium phosphates. *J. Am. Chem. Soc.*, 102, 2637-2643.

Roufosse A.H., Aue W.P., Roberts J.E., Glimcher M.J., Griffin R.G. (1984). Investigation of the mineral phases of bone by solid state phosphorus-31 magic angle spinning nuclear magnetic resonance. *Biochemistry*, 23, 6115-6120.

Rusu V.M., Ng C.H., Wilke M., Tiersch B., Fratzl P., Peter M.G. (2005). Size-controlled hydroxyapatite nanoparticles as self-organized organic-inorganic composite materials. *Biomaterials*, 26, 5414-5426.

Sarda S., Tofighi A., Hobatho M.C., Lee D., Rey C. (1999). Associations of low temperature apatites ceramics and proteins. *Phosphorus Res. Bull.*, 10, 208-213.

Scherrer P. (1918). Estimation of the size and internal structure of colloidal particles by means of R.overddot.ontgen rays. *Nachr. Ges. Wiss.*, Gottengen, 96-100.

Sfihi H., Rey C. (2002). 1-D and 2-D double heteronuclear magnetic resonance study of the local structure of type B carbonate fluoroapatite. In J. Fraissard, B. Lapina (Eds.) *Magnetic Resonance in Colloid and Interface Science, Nato ASI Series II*, pp. 409-418. Kluwer Academic Publishers.

Shirkhanzadeh M. (1994). X-ray diffraction and Fourier transform infrared analysis of nanophase apatite coatings prepared by electrocrystallization. *NanoStructured Materials*, 4, 677-684.

Sudarsanan K., Young R.A. (1969). Significant precision in crystal structural details: Holly Springs hydroxyapatite. *Acta. Cryst.*, B25, 1534-1543.

Suvorova E.I., Buffat P.A. (1999). Electron diffraction from micro-and nanoparticles of hydroxyapatite. *Journal of Microscopy*, 196, 46-58.

Tadic D., Beckmann F., Schwarz K., Epple M. (2004). A novel method to produce hydroxyapatite objects with interconnecting porosity that avoids sintering. *Biomaterials*, 25, 3335-3340.

Tampieri A., Celotti G., Landi E., Sandri M., Roveri N., Falini G. (2003). Biologically inspired synthesis of bone-like composite: self–assembled collagen fibers/hydroxyapatite nanocrystals. *J Biomed Mater Res*, 67, 618-625.

Termine J.D., Posner A.S. (1966). Infrared analysis of rat bone: age dependency of amorphous and crystalline fractions. *Science*, 153, 1523-1525.

Termine J.D., Eanes E.D., Conn K.M. (1980). Phosphoprotein modulation of apatite crystallization. *Calcified Tissue International*, 31, 247-251.

Thian E.S., Huang J., Vickers M.E., Best S.M., Barber Z.H., Bonfield W. (2006). Silicon-substituted hydroxyapatite (SiHA): a novel calcium phosphate coating for biomedical applications. *J. Mat. Sci.*, 41, 709-717.

Tropp J., Blumenthal N.C., Waugh J.S. (1983). Phosphorus NMR study of solid calcium phosphate. *J. Am. Chem. Soc.*, 105, 22-26.

Tsuboi S., Nakagaki H., Ishiguro K., Kondo K., Mukai M., Robinson C., Weatherell J.A. (1994). Magnesium distribution in human bone. *Calcified Tissue International*, 54, 34-37.

Wilson R.M., Elliott J.C., Dowker S.E.P., Rodriguez-Lorenzo L.M. (2005). Rietveld refinements and spectroscopic studies of the structure of Ca-deficient apatite. *Biomaterials*, 26, 1317-1327.

Winand L. (1961). Etude physico-chimique du phosphate tricalcique hydraté et de l'hydroxylapatite. Ann Chim (Paris) 13th series, 6, 951-967.

Wu Y., Glimcher M.J., Rey C., Ackerman J. (1994). A unique protonated phosphate group in bone mineral not present in synthetic calcium phosphates. *J Mol Biol*, 244, 423-435.

Yesinovski J.P., Eckert H. (1987). Hydrogen environments in calcium phosphates: ^{1}H MAS NMR at high spinning speeds. *J. Am. Chem. Soc.*, 109, 6274-6282.

Yesinovski J.P. (1998). Nuclear magnetic resonance spectroscopy of calcium phosphates. In Z. Amjad (Ed.), *Calcium phosphates in biological and industrial systems,* pp. 103-143. Kluwer Academic Publishers.

Yuan H., Li Y., de Bruijn J.D., de Groot K., Zhang X. (2000). Tissue responses of calcium phosphate cement: a study in dogs. *Biomaterials*, 21,1283-1290.

Zahidi E., Lebugle A., Bonel G. (1985). Sur une nouvelle classe de matériaux pour prothèses osseuses ou dentaires. *Bull Soc Chim Fr*, 4, 523-537.

INDEX

A

absorption, 14, 15, 51
absorption spectroscopy, 15, 68
achievement, 1
acid, 14, 42, 43, 47, 48, 56, 67
acidic, 13, 14, 42, 47, 48
activation, 54
adaptability, 6
additives, 61, 68
adhesion, 38, 40, 42, 45, 53, 61
adsorption, vii, 40, 41, 42, 43, 66, 72
Ag, 21
age, 3, 6, 74, 75
agents, 57
aging, 1, 5, 9, 11, 18, 33, 35, 39, 52, 59
aging population, 1
aid, 1
alanine, 43
albumin, 40, 56, 73
alcohol, 57
alkaline, 47, 48, 56, 60
alkalinity, 60
allografts, 1, 49
alpha, 42, 48
alternative, 14
amine, 43
amino, 42, 43, 66
amino acid, 42, 43
amino acids, 42

ammonia, 10
ammonium, 37
amorphous, 4, 5, 9, 16, 26, 47, 48, 53, 69, 71, 75
amorphous phases, 26
Amsterdam, 68
anabolic, 38
animals, 6, 60
anions, 27, 30
anisotropic, 72
anisotropy, 26, 56
antibiotic, 71
apatite, iii, v, vii, 1, 2, 3, 4, 5, 6, 7, 9, 10, 11, 13, 14, 15, 16, 17, 18, 19, 20, 21, 22, 23, 25, 26, 27, 29, 30, 31, 33, 34, 35, 36, 37, 38, 39, 40, 41, 42, 45, 46, 47, 48, 49, 50, 51, 52, 53, 54, 55, 56, 59, 60, 61, 62, 63, 65, 66, 67, 68, 70, 71, 72, 73, 75, 76
apatite layer, 53, 61, 62
apatites, vii, 1, 2, 3, 4, 5, 6, 9, 10, 11, 13, 14, 15, 16, 17, 18, 19, 20, 21, 23, 26, 27, 29, 30, 33, 35, 36, 37, 38, 39, 41, 42, 43, 45, 46, 49, 51, 52, 54, 59, 63, 65, 67, 68, 71, 73, 74, 75
application, 17, 54
aqueous solution, 54
aqueous solutions, 54
aqueous suspension, 56
arginine, 42
artificial, 70
ASI, 75

78 Index

assignment, 23
associations, 49, 56
atoms, 14
attachment, 42, 43, 61
attention, 16, 38, 40
availability, 31

B

basic fibroblast growth factor, 42
beams, 15
behavior, 2, 6, 29, 30, 38, 52, 56, 59, 60, 61, 65
beta, 72
bicarbonate, 11
binding, 41, 42, 55, 62
bioactive, 2, 45, 54, 63
biocompatibility, 49, 52, 55, 70
biodegradability, 55
biodegradable, 55, 73
biodegradation, 55, 56, 59, 60
biological, vii, 1, 2, 3, 4, 5, 6, 7, 9, 11, 14, 15, 18, 21, 23, 29, 30, 36, 38, 39, 45, 49, 50, 52, 53, 56, 60, 61, 63, 68, 69, 73, 74, 76
biological activity, vii, 38, 49, 50, 60, 61
biological behavior, 2, 52
biological systems, 2
biologically, 52, 53, 71
biomaterial, 2, 52, 53, 56, 61, 70
biomaterials, vii, 1, 2, 7, 24, 25, 35, 39, 42, 45, 46, 50, 59, 60, 62, 63, 68, 69, 71, 72, 73, 74
biomedical, 29, 48, 76
biomedical applications, 29, 76
biomimetic, 2, 7, 11, 42, 43, 47, 51, 52, 54, 55, 56, 60, 61, 62, 67, 71
biomineralization, 2, 40, 72
blocks, 46
blood, 39, 53
blood plasma, 53
body fluid, 5, 16, 36, 40, 50, 60, 61
body temperature, 48
bonding, vii, 7, 52, 53, 60, 62, 63
bone, vii, 1, 2, 3, 4, 5, 6, 7, 9, 10, 13, 14, 17, 18, 19, 26, 29, 30, 36, 38, 39, 41, 42, 45, 46, 49, 50, 52, 53, 54, 55, 56, 59, 60, 61, 62, 63, 65, 66, 67, 68, 69, 70, 71, 72, 73, 74, 75, 76
bone cement, 46
bone remodeling, 7, 39, 42, 60
bovine, 40, 68, 71, 73
Brazilian, 51

C

Ca^{2+}, 4, 6, 11, 30, 41
calcification, 55
calcium, 1, 3, 4, 5, 7, 9, 10, 11, 12, 13, 15, 16, 19, 20, 25, 27, 30, 36, 38, 39, 41, 43, 45, 47, 48, 49, 52, 53, 54, 55, 56, 65, 66, 67, 68, 69, 70, 71, 72, 73, 74, 75, 76
calcium carbonate, 3, 4, 47, 49
capacity, 37, 41
carbide, 1
carbon, 1, 66, 69
carbon dioxide, 69
carbonates, 6, 29, 33, 36, 37
carboxyl, 43
carboxyl groups, 43
carrier, 42, 70
cartilage, 65
casein, 56
casting, 56
cation, 27
cations, 6, 11, 30
cell, 14, 17, 19, 20, 38, 40, 42, 52, 56, 59, 61
cell adhesion, 38, 40, 42, 61
cell culture, 52
cell differentiation, 52
cellulose, 55
cement, 48, 60, 62, 66, 67, 68, 69, 71, 73, 76
ceramic, 7, 50, 54, 55, 56, 61
ceramics, vii, 1, 7, 45, 49, 50, 54, 55, 63, 70, 71, 72, 75
chemical, vii, 1, 2, 3, 5, 6, 7, 9, 11, 13, 14, 15, 20, 23, 25, 29, 31, 33, 34, 35, 37, 39, 42, 52, 54, 62, 63, 68, 74
chemical composition, 3, 5, 6, 9, 15, 33, 34, 35
chemical interaction, 31, 62

chemical properties, 1, 2, 11, 54, 74
chemistry, 67, 73
children, 39
chitin, 55
chitosan, 55, 71
chloride, 57
classes, 48
classical, 55
clinical, 52
clusters, 67
CO2, 11, 14, 56
coatings, vii, 1, 7, 16, 19, 45, 46, 52, 60, 63, 65, 68, 72, 75
cohesion, 51
cohesiveness, 45
collagen, 41, 54, 55, 67, 69, 70, 73, 75
colloidal particles, 75
compensation, 6, 20
competition, 41
compilation, 6
complementary, 2, 29
components, 2, 3, 26, 40, 56, 61
composite, 1, 7, 46, 55, 56, 70, 71, 72, 73, 75
composites, vii, 1, 18, 45, 46, 54, 55, 56, 63, 69, 72
composition, vii, 2, 3, 5, 6, 9, 10, 11, 13, 15, 17, 20, 30, 33, 34, 35, 52, 53, 56, 59, 60, 62, 63, 67, 71, 74
compositions, 5, 7, 9, 49, 67
compounds, 5, 15, 43, 59
compression, 51
concentration, 12, 13, 14, 15, 36, 37, 40
conductivity, 30
Congress, iv
conservation, 50
consolidation, 49, 50, 51
contaminants, 39
control, 10, 47, 48, 56, 72
controlled, 2, 7, 35, 53, 55, 56, 75
conversion, 5, 48, 71
copolymer, 56
correlation, 14, 20, 61
cortical, 6, 17, 19, 65, 71
CRC, 71, 74
cristallinity, 4

crystal, vii, 2, 4, 5, 16, 17, 18, 38, 40, 42, 45, 49, 50, 60, 61, 63, 65, 66, 68, 75
crystal growth, 5, 38, 40, 42, 50, 61, 65
crystal structure, 2
crystalline, 1, 4, 10, 15, 21, 26, 30, 46, 51, 54, 66, 67, 70, 73, 74, 75
crystallinity, 2, 5, 10, 18, 33, 51, 56, 65, 66
crystallites, 16
crystallization, 5, 71, 75
crystallographic, 19, 48
crystals, 3, 5, 6, 7, 13, 17, 25, 26, 29, 30, 41, 42, 43, 45, 50, 60, 61, 67, 70
culture, 52
curve-fitting, 14, 23, 24

D

decomposition, 9, 10, 14, 23, 24, 50, 52, 71
defects, 1
deficiency, 10
degradation, 50, 56, 59, 60
degradation rate, 56
degree, 10, 51, 68, 73
degree of crystallinity, 10, 51
delivery, 53
denatured, 55
density, 51, 66
dental implants, 52
deposition, 52
deposits, 53, 74
detection, 15
dielectric, 10, 43, 66
dielectric constant, 10, 43, 66
differentiation, 42, 52
diffraction, 3, 15, 16, 17, 18, 19, 20, 63, 66, 67, 69, 74, 75
diseases, 1
dispersion, 55
distribution, 69, 76
distribution function, 69
diversity, 3
dogs, 55, 70, 76
drug delivery, 53
drug release, 71
drugs, 53

dry, 24, 26
drying, 10, 24, 25, 29, 45, 50, 53, 56
duration, 42

E

electrical, 30, 50
electrical conductivity, 30
electrocrystallization, 19, 75
electron, 15, 27, 51
electron beam, 15
electron beams, 15
electron diffraction, 15
electron microscopy, 51
electronic, iv
electrostatic, iv, 41, 45
elongation, 18
embryonic, 6
energy, 9, 30
environment, 24, 26, 68, 74
epiphysis, 39
equilibrium, 13, 17, 61, 65, 68
equipment, 15
ester, 56
ethanol, 43, 66, 74
Europe, 7
evaporation, 56
evidence, 16
evolution, 5, 11, 15, 17, 18, 19, 33, 39, 42, 53, 67, 72, 74
EXAFS, 27
exothermic, 48
experimental condition, 19
expert, iv
extracellular, 60
extracellular matrix, 60
extrinsic, 45

F

fabrication, 70
femur, 17
fiber, 55
fibers, 55, 57, 75

fibrils, 69
fibroblast, 42, 72
fibroblast growth factor, 42, 72
fibrous tissue, 60
film, 70
films, 71
flexibility, 9, 55
fluid, 70
Fluorapatite, 70
fluoride, 10
fluoride ions, 10
Fourier, 14, 20, 74, 75
Fourier transform infrared spectroscopy, 74
Fox, 65
France, 65, 66, 68, 69
freeze-dried, 10, 11, 12
FTIR, vii, 14, 15, 20, 21, 23, 24, 25, 26, 29, 33, 34, 51, 63, 73
FTIR spectroscopy, 14, 15, 24, 33, 34, 51
fusion, 45, 49
FWHM, 17

G

Gaussian, 16
gel, 9, 10, 11, 19, 52, 54, 55, 70, 71, 72, 73
gelatin, 55, 71
gels, 11, 49, 50
glycine, 42, 66
gold, 70
grafts, 49
groups, 20, 23, 26, 36, 41, 42, 49, 55, 74
growth, 1, 5, 25, 38, 40, 42, 50, 53, 55, 61, 65, 66, 72
growth factor, 42, 53, 61, 72
growth factors, 42, 53, 61
growth rate, 5

H

HA, 3, 7, 16, 19, 21, 22, 42, 50, 52, 54, 55
half-life, 42
healing, 7, 42, 53, 68
health, 39

healthcare, 1
heat, 48, 49
heating, 13, 19, 20, 50, 51
heavy metal, 39
heavy metals, 39
heterogeneity, 2, 7, 17
heterogeneous, 53
high temperature, 7, 9, 49, 50, 52
histological, 69
homeostasis, 30, 36, 73
homogeneity, 52
homogeneous, 6
host, 55
hot pressing, 50, 51
human, 7, 19, 53, 55, 66, 68, 70, 73, 76
hybrid, 54
hydro, 9, 10, 53
hydrolysis, 9, 14, 48, 51, 53, 56
hydrophilic, 53
hydrothermal, 9
hydroxide, 19, 20, 26, 53, 61, 73
hydroxyapatite, 1, 3, 4, 5, 7, 16, 18, 19, 20, 21, 26, 41, 42, 51, 52, 53, 66, 68, 69, 70, 71, 72, 73, 74, 75, 76
hydroxyl, 25, 36, 67, 68, 74
hydroxyl groups, 25, 74
hydroxylapatite, 76
hydroxylation, 50
hypothesis, 3, 5

I

identification, 3, 4, 42
imitation, 1
immersion, 16, 36, 38, 53
immunological, 1
implants, 52, 53, 61, 62, 69
impregnation, 54
in situ, 46, 67
in vitro, 67, 69, 70, 74
in vivo, 55, 56, 61, 67, 69, 74
indirect measure, 13
induction, 61
industrial, 7, 53, 76
industrial processing, 53

infection, 53
inflammatory, 60
inflammatory response, 60
infrared (IR), 4, 21, 22, 23, 29, 37, 43, 72, 74, 75
infrared spectroscopy, 5, 74
inhibition, 65
inhibitor, 41
inhibitors, 5, 10, 40, 60
injury, iv
inorganic, 1, 13, 53, 65, 75
instability, 2
integration, 55
integrity, 55
interaction, 29, 41, 43, 45, 53, 73
interactions, vii, 2, 31, 42, 48, 50, 53, 54, 55, 59, 61, 62, 74
interface, 29, 36, 52, 62, 70
International Atomic Energy Agency, 70
intervention, 39
intoxication, 39
intrinsic, 52
ionic, vii, 12, 20, 24, 30, 37, 50, 53, 55, 63
ions, vii, 2, 4, 5, 6, 10, 11, 12, 13, 14, 15, 20, 21, 23, 26, 27, 30, 33, 36, 37, 38, 39, 42, 43, 45, 51, 53, 54, 55, 59, 61, 63, 66, 67, 68
isostatic pressing, 57
isotherms, 38

J

Japan, 70

L

labeling, 55
lactic acid, 56
Langmuir, 38, 70
lattice, 17, 19, 20, 36, 67
lattice parameters, 20
law, 16
leaching, 56
lead, 39, 42, 49
literature, 15, 18, 30, 55

location, 3, 39
London, 66
long distance, 61
low temperatures, 63

M

macromolecules, 45, 49, 63
magnesium, 5, 10, 11, 15, 38, 39, 54, 60, 66
magnetic, iv, 69, 73, 74, 75
magnetic resonance, 69, 73, 74, 75, 76
magnetic resonance spectroscopy, 76
maintenance, 1
MAS, 70, 76
materials science, 45, 50
mathematical, 14, 16
matrix, 31, 40, 50, 55, 60, 66, 74
matrix protein, 40
maturation, 2, 10, 11, 13, 15, 18, 25, 31, 33, 34, 35, 37, 38, 39, 40, 41, 45, 53, 60, 61, 67, 73
measurement, 50
measures, 14
mechanical, iv, 1, 7, 31, 49, 50, 51, 52, 54, 56, 62, 68
mechanical behavior, 56
mechanical properties, 7, 31, 49, 50, 51, 52, 54, 56, 68
media, vii, 5, 9, 24, 47, 53, 63
metabolite, 56
metals, 52, 71
Mexico, 68
Mg^{2+}, 11, 15, 30, 38, 65
mimicking, 2, 53
mineralization, 36, 40
mineralized, 1, 3, 4, 17, 40
minerals, 3, 5
mixing, 55
mobility, vii, 30, 50, 63
models, 65, 68
modulation, 75
molar ratio, 20, 33
molecular weight, 56
molecules, vii, 20, 26, 41, 42, 45, 52, 53, 54, 63

molybdenum, 13
morbidity, 1
morphology, 42

N

Na^+, 6
nanocomposites, 70, 71
nanocrystal, 7, 25, 39, 42, 48, 50, 59
nanocrystalline, vii, 2, 7, 9, 10, 11, 15, 16, 18, 19, 20, 21, 22, 26, 27, 29, 33, 34, 35, 36, 37, 38, 39, 40, 41, 45, 46, 49, 51, 52, 54, 56, 59, 61, 62, 63, 68, 70, 73, 74
nanocrystals, vii, 2, 5, 10, 15, 16, 17, 18, 20, 23, 24, 25, 26, 27, 29, 30, 33, 36, 38, 39, 40, 42, 45, 46, 48, 49, 50, 53, 54, 59, 60, 61, 62, 63, 75
nanomaterials, 29
nanoparticles, 75
natural, 19, 55, 56
natural polymers, 56
neoformation, 55
New York, iii, iv, 72
Nielsen, 74
nitride, 1
nuclear, 69, 73, 74, 75
nuclear magnetic resonance (NMR), vii, 20, 25, 26, 29, 63, 65, 66, 67, 69, 70, 72, 73, 74, 75, 76
nucleation, 40, 42, 53, 55, 61, 72
nuclei, 25

O

observations, 26, 42
optical, 3
optical properties, 3
organic, 9, 31, 54, 55, 62, 72, 74, 75
organic polymers, 72
organism, 56
organization, 70
orthopaedic, vii, 48, 49, 52, 61
orthophosphates, 30
osteoblasts, 60

osteoclasts, 70
osteogenesis imperfecta, 68
osteogenic, 42
osteoinductive, 7, 60, 71
osteons, 7, 60, 62
osteoporosis, 1, 38
oxide, 47, 53

P

parameter, 19, 35, 48
Paris, 1, 48, 76
particle shape, 16
particles, 48, 52, 60, 73, 75
PCA, 1, 2, 10, 11, 12, 13, 15, 25, 47, 54
peptides, 42
performance, vii, 1, 2, 53, 63
periodic, 20
periodic table, 20
PGA, 56
pH, 2, 9, 10, 11, 23, 41, 47, 48, 53, 56
pharmacological, 38
phenylalanine, 43
phosphate, 1, 3, 5, 7, 9, 10, 13, 14, 15, 20, 21, 23, 25, 26, 27, 30, 36, 37, 38, 41, 43, 45, 47, 48, 49, 50, 52, 53, 54, 55, 56, 59, 61, 65, 66, 67, 68, 69, 70, 71, 72, 73, 74, 76
phosphates, 5, 6, 7, 10, 25, 41, 47, 54, 55, 65, 66, 67, 68, 71, 75, 76
phosphoprotein, 69, 75
phosphorus, 13, 14, 69, 75
physico-chemical characteristics, 35, 39
physico-chemical properties, 2
physiological, 2, 10, 11, 23, 47, 48, 53
PLA, 56
plasma, 50, 52, 53, 68, 69, 70
platelet, 5, 16, 18
platelets, 5, 18, 54
play, 38, 40, 42, 43, 60, 61
PMMA, 48
poisoning, 39
polyester, 56
polyglycolic acid, 56
polymer, 7, 53, 55, 56, 72
polymer materials, 7
polymers, 49, 56, 71, 72
polysaccharide, 55
polyvinyl alcohol, 57
poor, 2, 5, 7, 35, 50, 56
pores, 54
porosity, 55, 56, 60, 61, 75
porous, 1, 46, 52, 54, 56, 63
porous materials, 52
powder, 19, 49, 51
powders, 50, 69
precipitation, 9, 10, 11, 13, 19, 53, 55, 56, 61, 74
preparation, iv, vii, 2, 9, 11, 35, 46, 55, 56
prevention, 53
procedures, 24
progressive, 5, 33, 37, 45, 53
proliferation, 42, 56, 60, 61
promote, 55, 61
promoter, 41
property, iv
prostheses, vii, 7, 52
prosthesis, 52
protein, 1, 40, 41, 42, 54, 55, 56, 72
protein binding, 41
proteins, 2, 40, 42, 55, 56, 61, 62, 67, 75
public, 1
public health, 1
pyrophosphate, 10, 13, 14, 60

Q

quasi-equilibrium, 61

R

radial distribution, 69
radius, 20, 39
Raman, vii, 20, 21, 22, 23
range, 9, 16, 26, 27, 29, 48, 49
rare earth, 6
rare earth elements, 6
rat, 18, 65, 66, 71, 75
reactivity, vii, 2, 35, 40, 45, 49, 50, 52, 53, 54, 61, 63, 74

reconstruction, 55, 62, 67
reduction, 30
regeneration, 2, 54, 71
regular, 19, 53, 60
regulation, 36
regulations, 31
rehydration, 45
rejection, 1
relaxation, 26
relevance, 6
remodeling, 7, 17, 39, 42, 60, 62
repair, 1, 7, 9, 39, 46, 67, 73
research, 1, 2, 56
research and development, 1
researchers, 1
reservoir, 31, 36
residues, 42, 60
resistance, 51
resolution, 18, 20, 74, 75
rings, 15, 45
risks, 1
room temperature, 2, 4, 9, 10, 11, 33, 51
Royal Society, 4, 37

S

salt, 36, 47, 48, 57
salts, 11
sample, 11, 14, 16, 18, 19, 36, 38, 74
saturation, 43
SBF, 50, 53
scaffold, 54, 56
scaffolds, 54, 56, 71, 72
scanning electron microscopy (SEM), 51, 54, 73
science, 45, 50
scientific, 49
scientists, vii, 63
self-organization, 70
sensitivity, 39
separation, 56
series, 72, 76
serine, 43, 66
serum, 40, 73
serum albumin, 40, 73

services, iv
shape, 5, 16, 18, 51, 67
shoulder, 25
shoulders, 23
signaling, 42
silicate, 55, 72
silicon, 20
similarity, 59
simulated body fluid, 16, 50, 70
sintering, 49, 50, 56, 68, 69, 70, 75
sites, 4, 6, 19, 30, 36, 38, 39, 41, 42, 43, 49, 55, 61, 70
sodium, 55, 57, 72
software, 23, 24
sol-gel, 9, 52, 70, 72
solid phase, 42, 74
solid solutions, 17
solid state, vii, 25, 26, 29, 63, 70, 73, 74, 75
solid-state, 26, 67
solubility, 17, 35, 59, 61, 65
solutions, 9, 10, 17, 53
solvent, 56
specialists, vii
species, vii, 3, 26, 33, 34, 36, 37, 42, 43, 60, 63
specific surface, 30, 52, 61
specificity, 50
spectra, 25, 26, 72
spectrophotometry, 13, 14, 69
spectroscopic methods, 20
spectroscopy, 5, 14, 15, 24, 33, 34, 51, 67, 68, 76
spectrum, 14, 21
spin, 67
sputtering, 52
SRS, 47
stability, 5, 35, 38
stoichiometry, 33
strain, 16, 17, 18, 67
strains, 50
strength, 62
stress, 51
strong interaction, vii, 29, 42, 55
strontium, 11, 12, 15, 38, 39, 70, 72, 74
structural defect, 19

structural defects, 19
structural modifications, 19, 42
substitutes, 2, 45
substitution, 2, 6, 7, 9, 20, 36, 38, 39, 68, 71, 72
substrates, 45, 53
supercritical, 56
surface area, 29, 30, 52, 61
surface chemistry, 73
surface energy, 30
surface layer, vii, 27, 36, 37, 45, 55, 63
surface modification, 7
surface properties, vii, 45, 50, 63
surface structure, 29
surface tension, 30
surgeons, 46
surgery, vii, 1
symmetry, 21
synchrotron, 74
synthesis, 9, 10, 19, 23, 33, 34, 35, 42, 60, 70, 75
synthetic, 1, 2, 4, 5, 9, 11, 14, 15, 17, 18, 19, 26, 29, 35, 42, 56, 65, 69, 71, 73, 74, 76
synthetic bone, 1
systematic, 53
systems, 2, 9, 76

T

TCP, 7, 20, 47, 48, 54, 56
technological, 7
teeth, 70
temperature, vii, 2, 4, 7, 9, 10, 11, 19, 33, 45, 48, 49, 50, 51, 52, 54, 56, 75
tensile, 51
tensile stress, 51
test procedure, 51
theoretical, 20
theory, 5, 20, 21
therapeutic, 72
thermal, 14, 71
thermal decomposition, 71
thermal treatment, 14
thermally induced phase separation, 56
thermodynamic, 48

three-dimensional, 54
time, 3, 9, 11, 13, 19, 26, 33, 34, 35, 37, 40, 41, 47, 51, 52, 53, 59, 61
tissue, vii, 1, 2, 3, 41, 52, 54, 55, 59, 60, 61, 62, 68, 71, 72
tissue engineering, 54, 55, 72
titanium, 1, 16, 20, 52, 53, 71
titration, 13
topological, 53
trace elements, 66, 70
transformation, 53
transitions, 27
transmission, 1
transparent, 50, 70
transport, 39
tumor, 1

U

uncertainty, 13
unstable compounds, 15
UV, 13

V

vacancies, 19, 33, 59
vacuum, 10, 11
values, 17, 18, 51
variability, 5, 17
variable, 3, 11, 17, 75
variation, 17, 20, 48
vertebrates, 1, 3, 5
vibrational, 20, 21, 23, 73
vibrational modes, 20, 21, 73
viscoelastic, 56
viscoelastic properties, 56

W

water, 10, 11, 12, 20, 26, 41, 45, 51, 56, 57, 66, 72
water-soluble, 57
wet, 24, 25, 26, 68, 69

X

XANES, 27
X-ray, 3, 15, 16, 17, 19, 20, 63, 66, 68, 69, 74, 75
X-ray absorption, 68

X-ray diffraction (XRD), 3, 4, 15, 16, 17, 18, 19, 20, 39, 51, 63, 66, 69, 73, 74, 75

Y

yield, 9